Springer Series on
SIGNALS AND COMMUNICATION TECHNOLOGY

Signals and Communication Technology

Thomas Hempel (Ed.)

Usability of Speech Dialog Systems

Listening to the Target Audience

With 14 Figures and 12 Tables

 Springer

Dr. Thomas Hempel
User Interface Design
Siemens Audiologische Technik GmbH
Software Portfolio & Product Management
Gebbertstr. 125
91058 Erlangen
Germany
thomas.hempel@siemens.com

ISBN 978-3-540-78342-8   e-ISBN 978-3-540-78343-5

DOI 10.1007/978-3-540-78343-5

Springer Series on Signals and Communication Technology ISSN 1860-4862

Library of Congress Control Number: 2008921123

Typesetting and production: le-tex publishing services oHG, Leipzig, Germany
Cover design: WMXDesign GmbH, Heidelberg

Printed on acid-free paper

9 8 7 6 5 4 3 2 1

springer.com

Content

List of Contributors

Caroline Clemens
Zentrum Mensch-Maschine-Systeme
(ZMMS)
Graduiertenkolleg Prometei
Berlin Institute of Technology
D-10587 Berlin, Germany

Klaus-Peter Engelbrecht
Quality and Usability Lab
Deutsche Telekom Laboratories
Berlin Institute of Technology
D-10587 Berlin, Germany

Peter Fröhlich
Forschungszentrum Telekommuni-
kation Wien (ftw.)
A-1220 Vienna, Austria

Paul Heisterkamp
Daimler AG, Group Research and
Advanced Engineering
Infotainment and Telematics
D-89013 Ulm, Germany

Thomas Hempel
Software Portfolio & Product Manage-
ment
Siemens Audiologische Technik
GmbH
D-91058 Erlangen, Germany

Ulrich Heute
Institute for Circuit and System Theory
Faculty of Engineering
Christian-Albrechts-University of Kiel
Kaiserstr. 2
D-24143 Kiel, Germany

Lu Huo
Institute for Circuit and System Theory
Faculty of Engineering
Christian-Albrechts-University of Kiel
Kaiserstr. 2
D-24143 Kiel, Germany

Wiebke Johannsen
T-Systems Enterprise Services GmbH
D-10589 Berlin, Germany

Jan Krebber
Sysopendigia
FI-00380 Helsinki, Finland

Sebastian Möller
Quality and Usability Lab
Deutsche Telekom Laboratories
Berlin Institute of Technology
D-10587 Berlin, Germany

Frank Oberle
T-Systems Enterprise Services GmbH
D-10589 Berlin, Germany

Rosa Pegam
Ümminger Str. 23
D-44892 Bochum, Germany

Michael Pucher
Forschungszentrum Telekommuni-
kation Wien (ftw.)
A-1220 Vienna, Austria

Roman Vilimek
User Interface Design Center
Corporate Technology
Siemens AG
D-81790 Munich, Germany

Carl-Frank Westermann
MetaDesign AG
Leibnizstraße 65
D-10629 Berlin, Germany

Chapter 1

The first contribution in this book is by Frank Oberle who gives an ample overview and introduction to the field. Based on his broad experience in the field of voice user interface design, he identifies the key elements for the design of a successful speech application. While we often find a "one fits all" approach in today's systems, Frank points out the needs for different target groups – well fitting the book's subtitle: "Listening to the Target Audience".

Dipl.-Ing. Frank Oberle

Frank Oberle works as a consultant on architecture and design in the field of innovative voice and multimodal solutions at Deutsche Telekom's system integration division T-Systems. He leads work in the field of innovative application design and is responsible for the creation of tools for speech application development at T-Systems. He has been involved in several Deutsche Telekom innovation projects with a special focus on the integration of new technologies into speech dialogue systems and the conception of a framework for multimodal solutions. He studied electrical engineering at the Technical University of Berlin and has longstanding experience in speech technology and the development of speech applications.

1 Who, Why and How Often? Key Elements for the Design of a Successful Speech Application Taking Account of the Target Groups

Frank Oberle
T-Systems Enterprise Service GmbH, Berlin, Germany

1.1 Introduction

Cost reduction is of increasing importance for medium and large enterprises. Seen in this context, Interactive Voice Response (IVR) systems are becoming more and more significant. IVR systems can help to automate business processes as for example in call centers, which are now a growing market for IVR systems.

Automatic speech recognition (ASR) is the key technology that IVR systems are based on. Ongoing developments in speech recognition, which started some years ago with the recognition of single words or very short phrases, mean that natural spoken speech can now be understood (Natural Language Understanding – NLU), even though this is still limited to a specific speech domain. Restricted to very simple command and control tasks with a very low level of complexity in the early days, today speech applications can be used in highly complex tasks such as messaging, office applications, travel and hotel booking, restaurant and city guides. In addition to cost pressures, the maturity of ASR technology has now resulted in a high level of market penetration. But what are the basic factors that make an IVR system really successful?

The quality of the IVR system is a vital factor whose importance was repeatedly underestimated in the past. To ensure that a communication can be performed automatically with a content user, the IVR system's overall quality has to meet the customer's expectations. If the user becomes annoyed, irritated or even angry about a low quality in the IVR system so that they break off the communication and hang up, the costs are fatal – a customer is lost. Very different parameters contribute to the IVR system's overall quality. The basic condition is of course a working, error-free and fully functional IVR system. Stability and performance has to be ensured for all components of the IVR system. Although this is a fundamental requirement for the successful operation, commercial success of an IVR system also requires acceptance by the customer. Parameters to measure the acceptance thus contribute to the IVR system's overall quality.

Because of the complexity of the development process, every single step has to be reviewed using specific quality tests, to ensure that the final result meets the customer's expectations. Section 1.2 will therefore outline a methodical procedural model to build a speech application that is properly designed for its target group, which covers the analysis of requirements, specification, implementation, production, delivery and operation. Particular attention will be paid here to the issue of quality assurance for IVR systems.

As mentioned above, commercial success of an IVR system also requires acceptance by the customer. To achieve this, the IVR system has to be aware of the customer's capabilities, needs and expectations. The speech dialogue has to be adapted to its target audience; it has to keep in mind who its customers are, what they want and how often they will use it. Expensive publicity campaigns may indeed help to acquire a large number of customers. However, the essential question is whether the customer will call again.

Section 1.3 will give an overview of the most important information we need to describe a voice user interface (VUI) suitable the target groups concerned and where we can get this information. This chapter also provides an overview on current and future technical developments in the field of speech processing and their relevance for the design of future dialogues.

To define some fundamental rules for the adaptation of dialogues, Sect. 1.4 will outline how to reduce the complexity of information to a simple, three-dimensional view of the caller, describing the user from the dialogue designer's perspective and focusing on characteristics for which the dialogue side can be modified.

Section 1.5 will discuss design features which enable the designer of a voice user interface to exploit knowledge about the user and to focus the design of the dialogue on the user's abilities, competence, expectation and needs.

1.2 Quality Assurance for the Development of IVR Systems – a Procedural Model

In this chapter a methodical procedural model will be outlined which describes the workflow to build a speech application suitable for a known target group, as used by T-Systems for the development of speech applications. The various project phases such as the analysis of requirements, specification, implementation, production, delivery and operation will be described, paying particular attention to the quality assurance issues. Despite the sequential presentation below, the procedural model described should of course be regarded as an iterative process.

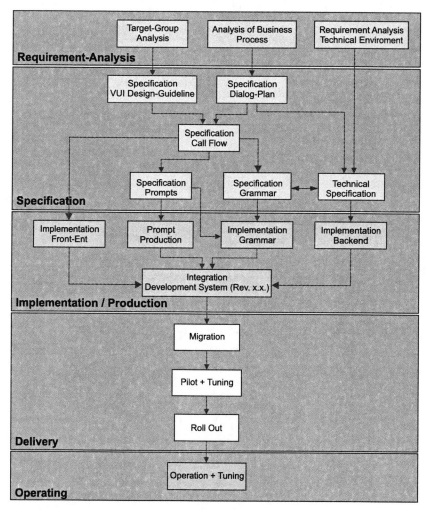

Fig. 1.1. Procedural model for the development of IVR-systems

1.2.1 Analysis of Requirements and Specification

The analysis of the requirements covers the target group, the business process and the technical environment, and must be done in close coordination with the customer for the IVR system.

When analyzing the target group we have to evaluate who will use the IVR system, why and how often. This analysis results in the definition of design guidelines describing the dialogue concepts and strategies (see also Sect. 1.5) which will be realized within the dialogue.

The objective of analyzing the business process is to identify the potential for automation within the process. As a result, the dialogue plan has to be defined, describing the rough structure of the speech dialogue (e. g. represented in the form of a task tree [Paterno et al. 1997]).

The analysis of technical requirements deals with all those issues that have to do with the integration of the IVR system into existing infrastructure; the integration with the phone system, the integration with the database as well as loading and security requirements. A technical specification results from this, defining the system architecture (definition and dimensioning of the modules and components), the interfaces (e. g. to databases or to the phone system) and the concepts for security, installation, configuration and operation (logging, reporting, monitoring).

A detailed specification of the dialogue has to be defined (most commonly represented as a call flow). This starts from the design guidelines and from the dialogue plan, specified within the analysis of the target group and the analysis of the business process. The specification of the dialogue is then the basis for the specification of the voice prompts and of the grammatical rules for the speech recognition. The specification of the prompts covers the formulation of the system output whereas the grammatical specification defines the slots (a slot is defined for each item of information that is relevant to the application) and appropriate return values for each dialogue state.

A review of the detailed specification of the dialogue and the system prompts by designated usability experts should contribute to the quality even at this early stage of development. Whereas for most of the projects a review by a usability expert may be sufficient, a Wizard of Oz test might be recommended for major projects [Fraser, Gilbert 1991]. The key idea of WOZ is to simulate the behavior of the working system by having a human act as the system. WOZ thus enables early testing even before a prototype exists.

1.2.2   Implementation and Production

The implementation and production phase covers the implementation of the front end (implementation of dialogue), the grammatical rules for speech recognition (implementation of the rule based ASR grammar files or the training of statistical N-gram approaches), the business logic, the back end as well as the production of prompts (i.e. the recording and processing of the sound files) and the integration of all components.

If the development environment allows "offline" simulation of the dialogue, e. g. text-based by typing the user utterance, it is possible to run traversal tests, which are aimed at checking how precisely the implementation conforms to the specification of the dialogue. Offline simulation of the dialogue allows the exclusion of all disturbing influences, e. g. caused by the speech recognition or by the prompts.

In parallel to this, the system prompts have to be reviewed in a separate test. Those tests are to ensure that the content of the prompts matches the specification and that the audio quality is satisfactory.

If rule based grammar is used, the first tuning of the ASR grammar should be done offline, using test sets. Test sets include a set of possible replies to a system prompt. The specification of the prompts has to be available for the test sets to be created.

Tuning the grammar offline includes coverage, NLU, over coverage and ambiguity tests. Coverage tests aim to test whether all varieties of meaningful formulations are covered by the ASR grammar. In addition, NLU tests have to verify that the varieties of formulations are represented by the appropriate semantic interpretation. It is recommended that test sets for coverage and NLU tests should not be created by the same person who implemented the grammar. Over coverage tests are meant to detect and eliminate all senseless and needless formulations. In general, the analysis will be based on test sets generated by a random process. Ambiguity tests are intended to detect and eliminate all cases in which one formulation is represented by more than one semantic interpretation.

Statistical language models (SLMs) are used when the amount of expected variation in spoken input is hard to capture with explicit grammar rules. When testing statistical language models, the most important concept is that the test set data must not be part of the training set. Tuning is done by adding new data sets of utterances, transcribed and labeled with the appropriate slot values.

After the integration of call flow, grammar, prompts and backend, integration tests have to check the interaction between the components.

After the integration of the single components, a first version of a complete and executable test system will be available. This represents the basis for the execution of the first expert and friendly user tests, where friendly users (e. g. colleagues) have to solve predefined test problems. On the basis of test protocols, log files and call recordings, first tuning of the entire system (call flow, prompts and ASR grammar) is possible.

Depending on the project budget available, the implementation and production phases have to be completed with evaluation by usability experts or with a usability test by users. These tests have to ensure that the IVR system fulfills the minimum requirements for the pilot phase. The basic condition for these tests is a working, error-free, fully functional test system. During evaluation by usability experts, as a rule three independent experts examine the system to find any usability problems from the user's point of view. The checks are based on previously defined usability requirements or on generally applicable usability criteria (e. g. standard ISO 9241-110/-11/-12 for the design of interactive systems [http://www.iso.org/]). A test for usability means that the system is checked by test persons instead of experts. On the basis of test problems, which should reflect representative usage of the system, test users are observed by an investigator, while interacting with the system. Based on the actions and reactions observed while interacting with the system, conclusions about the usability and the quality of the user interface can be drawn. A popular method here is to encourage the test person to think out loud, while interacting with the system. It is also usual practice to film the test person as they solve the problem. User tests of usability have to either be carried out in a usability laboratory or in a natural environment.

1.2.3 Delivery

After migration to the target environment, the stability and performance (e. g. response times) of the IVR system have to be tested with a performance or load test.

In a further step, an acceptance test is needed to ensure that the requirements for roll-out are met. To find any potential for optimization, the acceptance test verifies for example the perceived utility or the reasonableness of the price. For this reason, about 100 or 200 test persons should use the system, in a free and natural environment, following predefined instructions. Feedback is then collected via questionnaires or log files. Functionality, reliability, orientation, suggestibility, comfort and aesthetics are points to be considered and are parameters to measure the acceptance of IVR systems.

Based on the results of the pilot phase – that is log files, call recordings, acceptance tests or interviews of pilot attendees – the IVR system has to be tuned for one last time before roll out. During this phase, the tuning of the speech recognition is particularly time consuming. The speech recognition tuning covers the tuning of the speech recognition engine and, if applying rule-based grammar, the tuning of the dictionary, probability tuning and once again, grammar tuning.

When tuning the speech recognition engine itself, the classification rates for "correctly accepted", "incorrectly accepted" and "incorrectly rejected" have to be optimized by adjusting parameters such as the rejection threshold, the start of speech and the end of speech timeout or the pruning threshold. Prepared utterances will be presented to the engine in batch mode and the classification rates achieved for the various parameter settings will be compared. Dictionary tuning works by adding those unusual pronunciations to a dictionary, for which recognition errors occurred (e. g. the proper names of towns, companies or products). For grammar with a large vocabulary, significant improvement can be sometimes obtained by adding a priori probabilities to the grammar files, called probability tuning. Thanks to a final tuning of the grammar, frequently spoken utterances (which might cause an out of grammar event) have to be identified and added to the grammar. After tuning the grammar, coverage and NLU tests have to re-check the performance.

For statistical language models (SLMs) the new data sets of transcribed and labeled utterances, recorded during the pilot phase, have to be added to the training set. Because the data recorded during the pilot phase is more realistic than the old, laboratory-based data it might even be sensible to completely replace the old data set.

Finally, you should not forget that testing and tuning has to be carried on continuously, after roll-out and during operation of the IVR system.

1.3   How to Collect the Key Information for Design for a Target Group

As mentioned above, a working, error-free and fully functional IVR system is not the only essential factor for the commercial success of IVR systems. In order to get the necessary customer acceptance, the IVR system has to take the customer's capabilities, needs and expectations into consideration. Three important questions have thus to be answered:

- Who will use the system?
- Why will they use it?
- How often will they use it?

With answers to all of these questions, the designer will be able to best match the system to the target group, which will help to increase usability and efficiency. But where does the information come from?

In most cases, the questions "why" and "how often" are easy to answer. The application designer will exactly understand the business process the speech application should reproduce. We don't need to point here out that expectations and needs differ enormously; for example if the customer has to deal with an infotainment system or with a banking application. Furthermore, the business process allows us to draw conclusions about the frequency with which the customer will use the application. Will customers only use it two or three times a year, or almost daily, as for example with a messaging service? This is an important factor for the design of speech applications.

In personalized services with access to user data, the question "who is going to use the application?" is also easy to answer, because a lot of information about the user is available as soon as the caller has been identified by the system. In many services however, information about the caller is not available. In cases like this, the business process might allow us to draw some conclusions about the expected, average user. On the periphery of infotainment and games, a younger and open minded user can be assumed. Applications with a well defined target audience are however rare. Speech applications are mostly conceived for a complex, varied audience.

So, where do we get the information about the caller from?

A lot of information can be obtained from analysis of the CLI (Caller Line Identity, the presentation of the incoming caller's telephone number). If available, the area code provides information about the location of the caller. We can also determine whether the caller is using a mobile phone and is probably distracted. If, in addition, a database allows tracking data to be kept, the information can be used for the personalization of later calls.

The analysis of environmental information, such as the background noise, can provide information as to whether the caller is in a quiet or a loud environment or whether the caller is currently driving a car and is distracted. Global positioning and navigation systems such as GPS (global positioning system) are on the rise. Mobile devices equipped with GPS receivers allow the caller's location, speed, direction and time to be determined.

Information about the human machine interaction such as the frequency of certain events (interruption, no match, no input or help), the average response delay or the average number of words per user turn is easy to trace and can be extracted from log files.

Another possible solution might be the analysis of the speech signal, recorded from the caller. Approaches to the classification of speakers, based on the analysis of the speech signal, come under the term "Speaker Classification". Since speaker classification is key technology to close the knowledge gap on the dialogue designer's side, current research is increasingly addressing this subject, which will lead to an increase in performance in the near future.

1.3.1 Speaker Classification: Parameters and Methods

The approach here consists of extracting features, which are based on the analysis of the recorded speech signal from the user, and feeding them into a classifier. As this chapter is not intended to describe this approach in detail, the following section will simply give a compact overview of the state of the art, of the parameters and of the methods of determining them. Instead we will concentrate on rating of the degree of maturity of the approaches (see Sect. 1.3.2).

Speakers' characteristics can be differentiated by the degree of variability across time. Characteristics such as *gender, native language, dialect, accent, age* or *state of health* are more or less time invariant within a dialogue. These characteristics are immediately available, after the first user utterance, and can be used to modify the whole dialogue. While simultaneously supplying the user utterance in parallel to both a speech recognizer and a speaker classification module, the speaker classification does not restrict the progress of the dialogue, as no additional question is necessary. On the contrary, characteristics such as *emotion* (*joy, mourning, anger,* etc.), can rapidly change within an interaction. They thus have to be tracked continuously during the dialogue to ensure that appropriate changes to the dialogue are immediately initiated, e. g. transferring the caller to a human agent, if anger is detected.

In the first step, features are extracted to transform the digital speech signal into a parametric representation. Depending on the level of information, some features are better suited for certain classification criteria than others. Distinctions are thus drawn between spectral, prosodic, phonetic, idiolectal, dialogic and semantic features.

Acoustic features such as jitter, shimmer or the position of formants, are based on the spectral properties of small frames of speech of about 10–20 ms. Higher level features like pitch, duration, intensity or the duration of voiced and unvoiced sequences can be categorized as prosodic features. As an example, an anger detector, primarily based on prosodic features is described in [Burkhardt et al. 2005]. The probability that certain speech samples belong to a phonetic set, as well as the computation of pronunciation variants are called phonotactic features and seem to be especially useful to detect languages or dialects as well as emotion [Kienast et al. 1999]. Idiolectal features like the prediction of the use of certain words can be used to assign a speaker to a certain age or social group [Lee, Narayanan 2005].

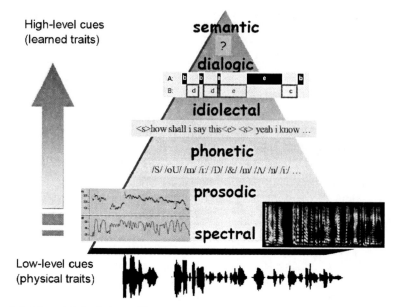

Fig. 1.2. Pictorial depiction of levels of information (from Reynolds et al. 2003)

Dialogic features comprehend all those features that have to do with human machine interaction such as the frequency of certain events (interruption, no match, no input or help), the average response delay or the average number of words per user turn (see also [Walker et al. 2002]). Semantic features can be used to detect the speaker's intention or to track the caller's preferences.

In a second step a classifier describes a set of discrimination functions aimed at mapping specific feature vectors to particular target classes. Because those approaches are based on a training set of the feature vectors, the performance of classifiers shows a strong dependency on the quality of the training set. The model extracted from the training set is then used to classify unknown data. Depending on the classification problem, e. g. classification of items with fixed or variable length or the realization of linear or non-linear discrimination problems, a variety of approaches is available (e. g. linear discriminant functions, dynamic Bayesian networks, K-nearest neighbour, Gaussian mixture model classifiers, artificial neural networks [Duda et al. 2000], support vector machines [Schölkopf, Smola 2002] or hidden Markov models [Rabiner 1989]). Depending on the complexity of the discrimination problem, they may also be used in combination, thus doing the pre-processing with one approach before applying another classifier.

1.3.2 Maturity of Speaker Classification

Experiments to evaluate different approaches to the classification of age and gender on telephone speech have shown: "We find that the best system performs on average comparably to human labelers, although the performance of our classifiers is worse on short utterances. [Metze et al. 2007]."

To what extent age is represented by the voice depends on so many individual factors, that the detection of age based exclusively on the analysis of the speech signal is even difficult for humans. Since there is reason to doubt that automatic classifiers will outperform humans instantly, the human results may be a good point of reference for the coming years: "Our results show that estimates of age are generally accurate only for broad age classes and confirm the intuitive expectation that people are able to assign a general chronological age category to a voice without seeing or knowing the talker [Cerrato et al. 2000]."

The results of an investigation with 7 age groups of 7 year ranges seemed to outline "that the listeners' performance is not very good" [Cerrato et al. 2000]. Forming 7 groups, the overall percentage of correct answers is on average only 27%. Forming only three groups (young, adult and old), the overall percentage of correct answers rises up to 71%.

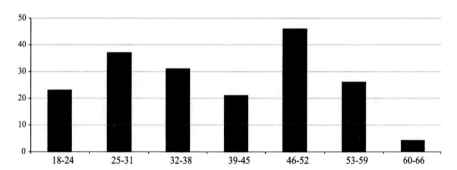

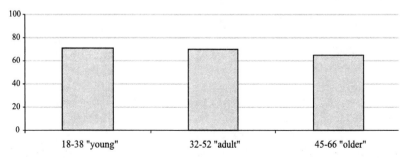

Fig. 1.3. Percentage of correct answers for 7 and 3 age classes (from [Cerrato et al. 2000])

Furthermore it must be noted that the human as well as the automatic results show a "centralization" trend for the perceived age, thus the phenomenon of overestimation of young speakers and underestimation of older speakers.

The differentiation between male and female here is very dependent on the age. Before boys' voices break, it is difficult to tell girls and boys apart by analyzing the speech signal. Differentiation between males on the one hand and females and children on the other hand is therefore easier than between females and children.

Since separate training of language models for males and females resulted in an improvement of recognition rates, gender detection is often carried out as an intermediate or preprocessing step for tasks such as speech recognition or speaker classification. Gender detection thus seems to be a mature technology, already offered by well established speech technology providers.

Anger detection, based on analysis of the recorded speech signal from the user, still seems to be a difficult task. First experience has shown that the encouraging results achieved under laboratory conditions still cannot be completely reproduced under real conditions. However, the anger detection already gives "results well above chance level" [Burkhardt et al. 2006]. Since the occurrence of false alarms still cannot be excluded, any resulting modifications to dialogue have to be selected carefully.

Another approach is the detection of anger based on a semantic analysis of speech. The occurrence of four-letter words may help. Also an analysis of the word order can lead to success. However insufficient experience has been collected so far.

Automatic language identification may help to pick up foreign language speakers before they hang up. By detecting the language of the caller, the IVR system will be able to switch the language or transfer the caller to a human agent who speaks their language. Seen in the context of an increasingly multicultural population and the effects of globalization, it seems more and more important not to exclude this part of the population. Multilingual services with automatic language identification might be the key.

Language identification can be accomplished as a by-product of speech recognition by feeding the utterance into grammar packages of different languages. If one of the packages maps to the utterance, the language is implicitly detected. If the utterance maps to more than one package, it is not easy to predict the language, because the confidence scores of different packages cannot be directly compared. The confidence score is a quantitative measure of how confident the recognizer is that it came up with the right hypothesis. Even more unsatisfying is the situation when no grammar package can map. In this case, determination of the language is impossible.

Thus, it is better to feed the user utterance into a parallel language identification module, to get a prediction that is completely independent of the speech recognition. Both results, i.e. the result of the language identification as well as the result of the speech recognition, can then be used, one result verifying the other.

Whereas language identification is almost ready for the market, the identification of accents, dialects or sociolects is just at the beginning.

Accent detection, able to provide the IVR system with hints about the native language of the caller would be particularly useful. The IVR system would be able to ask the caller, if they would prefer to continue in their native language.

The identification of dialects would be of particular interest, if language packages which are optimized for these dialects, to improve the performance of speech recognition, were available. As language packages that are optimized for special accents are not available (with the exception of e. g. U.S. and British English), initiatives in this direction are still limited.

Varieties of speech which are not based on regional but on social differences are called sociolects. Sociolects will be the key for an automated prediction of socio-cultural milieus. Current market research no longer only categorizes people depending on their social situation but also reflects their orientation. This has resulted in the definition of sociocultural milieus such as Sinus or the transnational Meta milieus (http://www.sinus-sociovision.de/), defined for Western Europe.

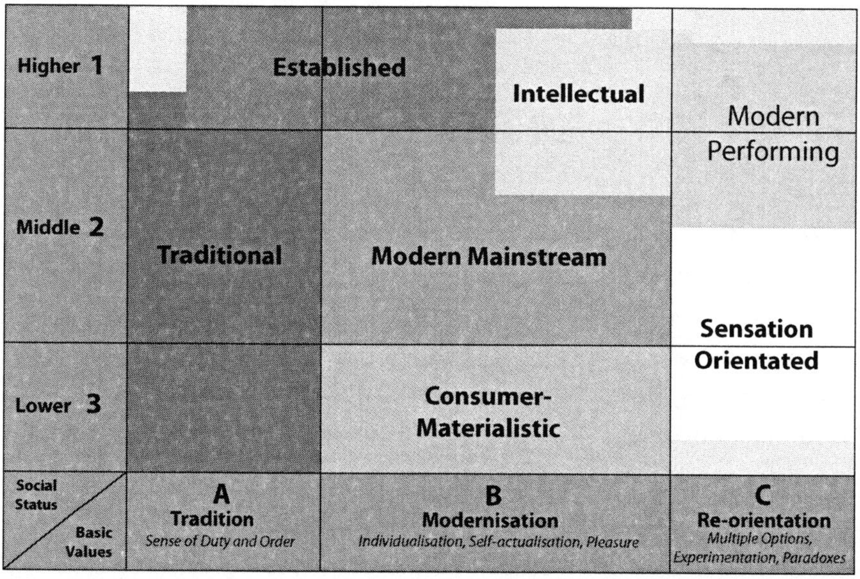

Social Status	Basic Values	A Tradition Sense of Duty and Order	B Modernisation Individualisation, Self-actualisation, Pleasure	C Re-orientation Multiple Options, Experimentation, Paradoxes
Higher 1		Established	Intellectual	Modern Performing
Middle 2		Traditional	Modern Mainstream	Sensation Orientated
Lower 3			Consumer-Materialistic	

Fig. 1.4. Meta-milieus (from [http://www.sinus-sociovision.de/])

If it were possible to predict a speaker's membership in a certain socio-cultural milieu on the basis of their speech signal, it would be possible to transfer all of the results and discoveries of modern market research at once into the world of IVR systems. Unfortunately, the research has not progressed that far and it is not an easy task to describe sociolects e. g. on the basis of the occurrence or frequency of specific words or by the position of words in a sentence.

1.4   How to Design Dialogues for a Target Group

In the above section we described how much information we can get about the user even when there is no user profile database available. But what is the relevance of this information for the dialogue designer? What are the changes to the dialogue that we recommend for the features identified?

In order to define some significant rules to change dialogue, first of all it is important to describe the user from the dialogue designer's point of view, focusing on characteristics for which the dialogue side can be directly modified.

Despite the fact that some of the following categories overlap and affect one other, they do reduce the complexity of information about the user to a simple, three-dimensional view of the caller:

- the user's ability to interact,
- the user's socio-demographic characteristics and
- the user's linguistic characteristics.

1.4.1   How to Consider the Caller's Ability to Interact

First of all, it is the business process that allows us to draw the most important conclusions. If it can be assumed that the customer will use the application only two or three times a year, we have to build the application for a caller who is most probably new and not familiar with the application. If we have to assume, that the application will be used with a high frequency, e. g. a messaging application, which will be used two or three times a day, we also have to offer dialogue strategies, which take power users into consideration, who are possibly very familiar with the application and who want to avoid wasting too much time in the application.

In a second step, we have to consider the affinity of people to information technology, whereby a strong dependency on age and gender has to be taken into account.

For the domain of science and technology the differences which depend on gender are so obvious, that there is talk of a "technological gender gap" ([Canada, Brusca 1992], [Anastasi 1976], [Asendorpf 2003]). Even schoolboys, but also male students are significantly more interested, are more self-confident about their own abilities and more often have their own computer or at least have experiences with a computer ([Sproull et al. 1984], [Levin, Gordon 1989]). Transferred to the domain of IVR systems, it is still hard to define gender related differences when designing dialogues. The only difference which might be assumed is that, on average, males will have less fear of contact with new technology and will have a greater motivation when learning to interact with it. Designers might thus consider how to introduce females more carefully to the interaction with IVR systems.

In terms of the age of a caller the differences are even more obvious. Elderly people seem to have increasing difficulty keeping up with the growing level of technology in everyday life, while the preparedness to accept change decreases at the same time ([Strong 1943], [Rudinger 1994], [Gilly, Zeithaml 1985]). This phenomenon is due to the substantial limitations elderly people are subject to, which has to do with their sensory abilities (speed of perception, ability to see and hear) and cognitive abilities (speed of answering, memory and ability to learn). Newer studies in developmental psychology have shown that the speed of reception and handling of information decreases rapidly with increasing age [Baltes 1990]. Learning the interaction with new computer technology is thus harder for elderly people [Gomez et al. 1986]. This has to be considered when designing interfaces for senior citizens.

On the technical side of this topic, experience in applying speech recognition has shown that a negative impact on the performance of speech recognition can be observed:

- for inarticulate speech – the probability of this increases for very young and very old callers,
- for over-long and freely formulated sentences – experience has shown that people who have no experience in interacting with IVR systems are particularly prone to this. As studies have shown, this particularly applies to the group of very young and old speakers ([Hempel 2006a], [Hempel 2006b]),
- for over-loud or over-emphasized speech, which applies to callers who are exited or angry,
- for non-native speakers (with a strong accent) or for speakers with a strong dialect or who use heavy slang or sociolect,
- for a noisy environment.

Summarizing, it can be stated that a negative impact on the interaction between human and IVR system can be assumed and changes to the dialogue are recommended

- if a low frequency of usage is to be expected and the vast majority of users will be new and not familiar with the IVR system,
- for very old or very young callers,
- for exited or angry callers,
- for non-native speakers (with a strong accent) or for speakers with a strong dialect or using heavy slang or sociolect,
- for a high frequency of the speech recognition events "no input" and "no match" or for long response times of the human,
- if a noisy environment (e. g. railway station or car) or high distraction (e. g. call from a car or call from a mobile phone) can be assumed.

To estimate the caller's ability to interact, very different parameters can be considered, depending on their availability. The more parameters that support a specific hypothesis, the more likely this hypothesis can be considered.

To prevent callers from not being able to cope and to reflect the caller's ability to interact, appropriate changes to the dialogue might affect:

- the balance between mixed initiative and directed dialogue,
- the usage of confirmation and feedback strategies,
- the design of menu trees (depth of the tree structure, width of the menus),
- the setting of speech volume, speed and pauses,
- the highlighting of keywords in the user guidance,
- the use of touch-tone or the offer of assistance from a human operator.

1.4.2   How to Fulfill the Caller's Expectations and Needs

If the objective is to meet the callers' expectations and needs, by providing an interface tailored to a target group, the focus has to be on a socio-demographic classification of the caller.

Socio-demographic caller classification can be based on age and gender as well as on the caller's membership of certain socio-cultural milieus, obtained from a user profile database or the analysis of spoken utterances. Since the prediction of membership in a socio-demographic milieu, based on the analysis of the speech, is not currently possible, for the moment, speaker classification has to be exclusively based on the age and gender of the speaker.

To meet callers' expectations and needs and to effectively tailor for target groups, dialogue modifications might consist of:

- consideration of target group preferences when defining the order and hierarchy with which themes and content are presented – target group preferences can result from market research or can be measured,
- design of the system's persona and voice (e. g. pitch, intonation, vocabulary, usage of technical terms and colloquial speech,....),
- consideration of target group preferences when defining background music, teasers, jingles or advertisements (e. g. when waiting in a queue).

But what are the specific preferences of males and females and of specific age groups? Behavior during leisure time and the use of media seem to be very different, depending on age and gender.

The annual studies AWA [http://www.awa-online.de/] and ACTA [http://www.acta-online.de/] by IfD Allensbach, in Germany, show that with increasing age, interest significantly recedes for consumer electronics, photography, video and for activities such as cinema, discotheques and computers. According to the results of the "Media-Analyse 2004/II" study carried out by the German television station "SWR" [SWR 2004], interest for activities at home or in the immediate neighborhood gains in importance with increasing age. This preference of older people for activities such as reading, watching TV, handicrafts or knitting can be explained by their decreasing mobility.

The differences between males and females are vaguer, as shown by the annual studies by "IconKids & Youth" [http://www.iconkids.com/] and by the German television station "SWR" [SWR 2004]. Males seem to prefer subjects like sports, cars or going out. Females, in contrast, prefer literature, shopping or cosmetics. Within the teenage target group, girls prefer fashion, shopping and cosmetics, whereas boys prefer sports, cars and computers.

Gender related differences in the evaluation of anthropomorphic interfaces and the design of artificial agents are hard to verify. The results of different studies on this subject are too diverse to make conclusions about concrete recommendations for gender-specific design ([Bickmore, Cassell 2005], [Catrambone et al. 2004], [Buisine et al. 2004], [McBreen 2002]). At most we could take with the statement that females react more sensitively to aspects of artificiality than males ([Kraemer 2001], [Kraemer 2002], [Sproull et al. 1996]). In general, it can be summarized that apart from gender, acceptance always rests on a balance between naturalness on the one hand and transparency and functionality on the other one.

Although there are no substantial differences between males and females, both having the same vocabulary available, gender-specific usage of language does become apparent. Whereas females interact politely and cooperatively, using standard or high-level language with mitigating and understating formulations, males rather have a tendency to a colloquial and direct brusqueness with an imperative, emphatic and competitive conversational style (see also [Günthner 1997], [Braun 2004], [Mulac 1999]). This knowledge can be applied to realize a more authentic design of persona, in which a female artificial agent uses the typically female mode of speaking. This knowledge can also be used to meet the caller with the appropriate mode of speaking, where the IVR system is able to reflect the conversational style of the caller.

1.4.3 How to Consider the User's Linguistic Background

Native language, accent, dialect, sociolect and speed of speech contribute to the caller's linguistic characteristics. The results of a linguistic characterization of the caller might be reflected in following changes:

* switching the system prompts and the speech recognition package to the caller's native language,
* transferring the caller to a human agent who speaks their native language,
* improvement of speech recognition by switching to a speech recognition package that is optimized for a particular dialect,
* alter the system prompts to the caller's mode of speaking (speed of speech, dialect, sociolect).

1.5   Eleven Recommendations for Design for a Target Group

1.5.1 Mixed Initiative vs. Directed Dialog

In general there is a difference between mixed initiative and directed dialogues. If we take a couple of statements within an interaction as a basis for our observation, there is a very simple model of interaction if we define the first speaker as the acting speaker and the second speaker as the reacting speaker – the speaker who asks has the initiative and directs the interaction.

In directed dialogues the IVR system keeps the initiative during the whole dialogue. Directed dialogues are based on a conventional menu structure, the system's prompting emphasizes the keywords and the speech recognition is focused on the recognition of single words or short phrases. Directed dialogues need far less time and effort to implement.

Mixed initiative, on the other hand, enables the caller to take the initiative during the dialogue, whenever they want. The IVR system takes the initiative only if problems occur or if the dialogue is deadlocked. The caller is able to switch between several tasks of an application without using a menu structure. As it is usual in natural interaction, one user utterance is allowed to contain multiple items of information. The dialogue system should be able to check the completeness and validity of the information provided by the user utterance, and ask for any missing or invalid information. Mixed initiative requires speech recognition providing "Natural Language Understanding" (NLU) and "Multi-Slot Grammar", which are able to return more than one slot (i.e. more than one item of information) per utterance.

Example:

> Caller: I want to know the trains from Cologne to Berlin tomorrow.
> *IVR: Please tell me the time of departure.*
> Caller: Between 6 and 7 a.m.
> *IVR: Tomorrow between 6 and 7 a.m. I can offer the following trains:...*

Mixed initiative is accompanied by a lot of other enhanced dialogue concepts like overloaded answers, interruptability or ellipsis.

Answers to a system prompt are overloaded, if a user's utterance contains information which exceeds the question actually asked by the system. Storing this information for later usage in the dialogue allows the dialogue to be significantly shortened. More than one subsequent dialogue step can be often skipped if the additional information already defines a special subtask in the task tree.

Without overloading the answer, a travel booking dialog would look like this:

> *IVR: Do you need a flight ticket, a rail ticket or a hotel reservation?*
> Caller: Flight ticket.
> *IVR: Your point of departure?*
> Caller: Cologne.
> *IVR: And your destination?*
> Caller: Berlin.
> *IVR: Please tell me the day of departure?*
> Caller: Tomorrow.
> *IVR: And the time of departure?*
> Caller: Between 6 and 7 a.m.

When an answer is overloaded, the dialogue can be shortened, if the dialogue control is able to handle the additional information. In the following example the option "hotel reservation" can be excluded. The dialogue control only has to resolve the ambiguity between "travel by airplane" and "travel by railway". Furthermore the query for "destination" as well as the query for "day of departure" can be skipped:

> *IVR: Do you need a flight ticket, a rail ticket or a hotel reservation?*
> Caller: Tomorrow I want to travel to Berlin.
> *IVR: By airplane or railway?*
> Caller: Airplane.
> *IVR: Your point of departure?*
> Caller: Cologne.
> *IVR: And the time of departure?*
> Caller: Between 6 and 7 a.m.

Interruptability allows the caller to interrupt a task and initiate a new task whenever they want. When the new task is finished, the dialogue system is able to continue the previous task from the point where the task was interrupted.

In the following example the information "tomorrow" concerning the weather task is stored and re-used for the booking task:

> *IVR: Do you need a flight ticket, a rail ticket or a hotel reservation?*
> Caller: Flight ticket.
> *IVR: Your point of departure?*
> Caller: Cologne.
> *IVR: And your destination?*
> Caller: Berlin.
> Caller: How is the weather tomorrow in Berlin?
> *IVR: Tomorrow the weather in Berlin is rainy and cold. Take an umbrella and warm clothing!*
> *Let's continue with the booking of your flight. The day of departure is tomorrow?*
> Caller: Yes.
> *IVR: And the time of departure?*
> Caller: Between 6 and 7 a.m.

Elliptical expressions are if a caller provides just part of the information within the current utterance. In that case the IVR system needs to integrate it into its preceding context in order to infer its complete meaning. Elliptical expressions are very common in natural interaction. In the following example the system needs to integrate "flights" from the second user utterance into the context of the previous utterance, where "trains" has to be replaced by "flights" in order to completely understand the user's concern:

> Caller: I want to know the trains from Cologne to Berlin tomorrow between 6 and 7 a.m.
> *IVR: Tomorrow between 6 and 7 a.m. I can offer the following trains:...*
> Caller: And now I want to know the flights.
> *IVR: Tomorrow between 6 and 7 a.m. I can offer the following flights:...*

Since mixed initiative is based on an interaction between the IVR system and the caller at a very high level, it requires that the caller be very familiar with comparable systems. Directed dialogues thus seem to much better suit a caller who isn't familiar with the IVR system. Those callers definitely prefer directed dialogue to mixed initiative, if of course the dialogue is well designed.

This topic was investigated by "Siemens AG, Corporate Technology, User Interface Design Center" using a Wizard of Oz test and a usability test with 25 test persons each ([Hempel 2006a], [Hempel 2006b]). The usability test was based on a prototype T-Systems speech application, offering different dialogue strategies for different user groups. The 25 test persons, who attended the usability test, represented 5 different user groups (child, adult male, adult female, older male, older female).

Its results showed that directed dialogue is preferred by the test persons, even though it is far less convenient, when compared to mixed initiative. This lack of acceptance for mixed initiative in the first instance is due to

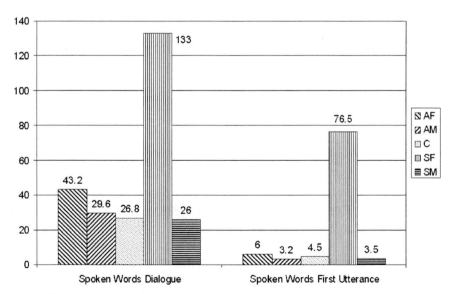

Fig. 1.5. Number of words uttered per dialog in the first user utterance (AF/AM = adult females/males; SF/SM = senior females/males; C = Teenagers) (from Hempel 2006a)

the test persons not being familiar with comparable systems. The acceptance for directed dialogues is even more pronounced for children and older people. Especially females and older people show a tendency to answer more verbosely. When testing mixed initiative they apparently accept the offer of a natural dialogue.

To take target groups into account when defining the balance between mixed initiative and directed dialogue, the following can be recommended:

- It can be assumed that people with a limited ability to interact with the IVR system, e. g. elderly people, non-native speakers or children, will prefer a directed dialogue. If there are signs of an impaired interaction, directed dialogue is recommended.
- For speech applications which will not be used very often, a directed dialogue is able to fulfill the requirements.
- For those applications which will be used very often, a combination of mixed initiative and directed dialogue is recommended. In this case the IVR system basically provides the option for the user to take the initiative and only takes the initiative itself if problems occur or the dialogue is deadlocked.
- A lack of acceptance for mixed initiative is primarily due to the caller's lack of familiarity with comparable systems.
- Open dialogues with mixed initiative cause very different expectations, depending on the age and gender of the caller.
- Even though mixed initiative is more convenient for the caller, it will not be accepted if the caller isn't aware of it. To become familiar with it as soon as possible, mixed initiative must be explained to the user.
- An explanation on how the interaction with IVR systems works might be particularly useful for older callers, even if only for directed dialogues. This can be done for example by presenting a short example dialogue.

1.5.2 Strategies for Feedback and Confirmation

Depending on the business process it can be more or less important to be sure about the words the caller has said. The system reacting incorrectly in an infotainment dialogue is in no way as hard to cope with as a comparable incorrect reaction within a banking application.

To be sure about the caller's intention, the system has to get confirmation from the caller. Confirmation can be given within an explicit or implicit confirmation state.

Explicit confirmation provides an explicit dialog state for confirmation:

> *IVR: And your destination?*
> Caller: Berlin.
> *IVR: Your destination is Berlin?*
> Caller: Correct.

To speed up the dialogue, the analysis of confidence scores may help. Most commercial speech recognizers return them in addition to the recognition hypothesis. The confidence score is a quantitative measure of how confident the recognizer is that it came up with the right hypothesis. Confidence analysis enables the IVR system to confirm only uncertain speech recognizer hypotheses with a low confidence score. However for safety critical and security sensitive business processes, an explicit confirmation state should always be planned.

"N best" enables access not only to the speech recognition result with the highest confidence score, but also to the N best results. "N best" allows efficient handling of confusion for particularly large vocabularies as for example all of the towns in the U.S.:

> *IVR: And your destination?*
> Caller: Austin.
> *IVR: Your destination is Boston?*
> Caller: No.
> *IVR: Your destination is Austin?*
> Caller: Yes.

If the caller denied Boston, the IVR system can be designed to pass over Boston and select Austin later in the dialogue, thus not repeating the mistake. Together with mixed initiative, the confirmation states can even reach a higher level of complexity. "One step negation and correction" enables the user to refuse and adjust a detail in one single dialogue step.

> *IVR: And your destination?*
> Caller: Austin.
> *IVR: Your destination is Boston?*
> Caller: No, Austin.

"Multi-Slot-Confirmation" allows the system to confirm more than one piece of information within one dialogue step:

> *IVR: I resume – you need a flight from Berlin to Cologne for tomorrow between 6 and 7 a.m.?*
> Caller: No, from Cologne to Berlin.

The most efficient way of speeding up the dialogue is to apply implicit feedback. The caller gets implicit feedback to his utterance within the next dialogue step, but without an explicit confirmation state. The handling of implicit feedback demands very complex dialogue management, since the following dialogue step has not only to handle the next task, but also the confirmation of the previous dialogue step:

IVR: Do you need a flight ticket, a rail ticket or a hotel reservation?
Caller: Flight ticket.
IVR: Your point of departure for your flight?
Caller: No, I want to take a train.

For usage of feedback and confirmation appropriate to the target group, the following strategies can be recommended:

- The application context is the most important issue for deciding upon the appropriate feedback and confirmation strategy. The confidence threshold for the activation of an explicit confirmation should first of all be adapted to the safety requirements of the actual business process. For absolutely safety critical and security sensitive business processes, such as transactions in a banking environment, an explicit confirmation state is a must.
- If no explicit confirmation state was activated in the previous dialogue state, at least implicit feedback should be given to the caller – at all times during the dialogue the caller should be aware about the status of the discourse.
- Since explicit confirmation doesn't place high demands on the caller's ability to interact with the dialogue system, this kind of approach seems to be particularly suitable for callers with a limited ability to interact (e. g. elderly people, non-native speakers or children) or for those speech applications that will not be used very often and where callers will not be very familiar with comparable systems.
- Because implicit feedback requires that the caller has a very high ability to interact with the dialogue system, it doesn't seem to be suitable for callers with a limited ability to interact or for those callers who are not familiar with comparable systems. Implicit feedback should thus be used carefully in those applications which presumably will not be used very often.
- It is therefore recommended that the confidence threshold for the activation of explicit confirmation should be adjusted not just to the safety requirements of the current business process, but also to the caller's ability to interact with the IVR system.

1.5.3 Structuring the Menu Tree

Although for mixed initiative dialogues the caller is able to switch between several tasks of an application without thinking about it, the menu structure should still be in the background as a fallback. This is in case problems occur in the interaction or if the dialogue becomes deadlocked. Directed dialogues are based on a menu structure anyway.

When structuring a menu tree, some elementary rules have to be considered. No more than three options should be presented within one menu, details should be in submenus and associated tasks should be grouped together within an appropriate menu. The caller should not go through more than three dialogue steps and not more than one minute should be necessary to reach their target.

It now seems that, the more complex the structure of an application, the more inconsistent the rules are with one another.

When designing the menu tree, the designer has to be aware of the user the application is being designed for. Some general rules can be mentioned for this:

- When designing a menu tree for users with an above average ability to interact with the dialogue system or for a speech application that will be used very often and where callers will be very familiar with the system, it seems sensible to optimize the length of the dialogue at the cost of the width of the menus. Also more than three options can be presented within a menu.
- When designing for users with a limited ability to interact with the dialogue system (e. g. elderly people, non-native speakers or children) or for speech applications that will not be used very often and where callers will not be very familiar with the system, the width of the menus should be optimized at the expense of the length of the dialogue. In this case a menu should never exceed 3 options.
- If signs are detected that the interaction between system and user is limited, the system should hand over to a human agent. If not available, it might be useful to verify the interest for each option by posing separate questions ("Are you interested in...?").

Regarding the order, in which the tasks have to be presented within a menu, popular tasks should be presented first and fall-back options like "abort", "main menu" or the option of a human agent should be presented last. It is thus useful to measure the frequency of usage of options in a menu. If preference for a certain option is very high, e. g. more than 80% percent of all requests in the menu, it might be useful to first verify interest for this option ("You are interested in...?") within a particular step in the dialogue before presenting the other options. In this way the dialogue can be simplified and distinctly shortened for most callers.

In this context, personalization is easy if the user's preferences have already been determined and saved during past calls. It then will be possible to first present the options that the user last requested. In general the question as to whether the advantages of such a practice outweigh the drawbacks, caused by the order in the menus permanently changing, possibly confusing the caller, is still controversial.

It thus seems to be useful to measure the correlation between the user's preferences and the user's socio-demographic parameters e. g. age and gender, obtained from a customer database or from speaker classification. The preferences specific to the target group then can be taken into account when structuring the menu tree. Compared to the nervy switching of the order when just considering the preferences of a single person, the above mentioned approach seems to be the better way. In this mode, by the way, marketing strategies can be directed exactly at the needs and expectations of the relevant target group.

For callers with a limited ability to interact with the IVR system (e. g. older people, children, non-native speakers,...) or if there are signs of impaired interaction, a human agent should be offered more directly, if available.

If there is information about the caller's location, speed, direction and time, e. g. obtained from a mobile device that is equipped with a GPS receiver, the offering can be tailored to the caller's specific situation. Options with particular relevance to the caller's current location and time can be presented first.

1.5.4 Escalation- and Help-Strategies

As mentioned above, a human operator should always be offered to resolve a deadlock in the interaction between human and machine and to make sure that the caller will not be lost. If only a limited number of human operators or none at all are available, escalation and help strategies have to get the caller out of trouble.

Another option is to offer DTMF (dual tone multiple frequency) control to overcome the user's difficulties with speech recognition. DTMF control still has a role anyway, especially in a noisy environment or if personal data has to be entered in a public environment as for example when entering a PIN (personal identification number) in a subway.

To avoid annoying repetitions and to guarantee progress in the interaction, multi-level escalation has to be offered to the caller. The relevant keywords, representing the options of the current dialogue state, and some essential hard keys such as "help", "abort" or "main menu" have to be emphasized. The "help" option as well as the fall-back commands "abort" and "main menu" have to be available throughout the application. The "help" option should offer some additional, specific information about the options in the current dialogue state. Commands such as "abort" or "main menu" should offer a way out of the current situation, back to a well defined point in the dialogue.

Even if escalation is not successful, the progress of the interaction should not be blocked. Again, information about the user's preferences may be useful in a case like this. Starting with the most popular option in a dialogue state, the relevance for the caller may be verified in an explicit question (e. g. "Are you interested in...?"). Alternatively, the system may automatically jump back to a well defined state of the interaction (e. g. back to the main menu), but not without a detailed comment such as: "I am sorry, I didn't understand again. Perhaps it would be better to start again. The current process will be aborted. Let's go back to the main menu."

In order to solve possible blocks in the interaction between human and machine, it is again useful to measure preferences and to correlate them with socio-demographic parameters such as age and gender, obtained from a customer database or from speaker classification. It then will be possible for the dialogue management to make a more precise prediction of the user's concern. In this case, statistical scores about the frequency of usage no longer have to be calculated using all potential users but can be defined for specific user groups. Beginning with the most popular option for the current user group, the option's relevance for the caller may be specifically verified ("You are interested in...?").

1.5.5 Degree of Automation

Call center automation is a growing market for IVR systems. In this environment, the main concern is to narrow down the user task through an automated speech dialogue and then to hand over to an agent who is particularly skilled in addressing the caller's wishes. The minimum requirement for the automated pre-classification is thus to roughly predefine the routing destination, i.e. the agent, who has the appropriate skill. Furthermore the automated dialogue can attempt to capture additional information from the caller in order to ease the agent's job.

In this context, the process can be optimized by adapting the degree of automation to the user groups' ability to interact with the IVR system. For user groups with a limited ability to interact, e. g. elderly people, the degree of automation can be restricted to the absolutely minimum, whereas for user groups with a proven affinity to information technology and where problem-free interaction with the machine can be assumed, e. g. male adults, an attempt can be made to shift the dialogue away from the agent and towards the automated dialogue.

1.5.6 Politeness-, Motivation- and De-Escalation-Strategies

As a response to the detection of anger, obtained from an emotion detector, the IVR system should immediately answer by offering a human agent. If not available, motivation, politeness and de-escalation strategies might help.

Politeness strategies are conceived to prevent negative emotional conditions such as dissatisfaction, stress, frustration and anger. Positive politeness aims to identify with the caller's objectives and to emphasize common interests in the interaction between human and machine. Negative politeness, instead, aims to relieve a caller under pressure and to eliminate all tension that may be caused by the interaction. This can be achieved by emphasizing that the caller is the one who makes the ultimate decision and that they can take as much time as they want.

Motivation strategies are especially useful for long dialogues, where the caller has to answer up to ten or more questions. A caller can be motivated by information about the progress already made and by encouraging the caller to carry on.

De-escalation strategies aim to divert the caller's attention away from the reason for dissatisfaction, stress, frustration and anger. This can be achieved for example by pointing out alternatives to the current situation.

1.5.7 Wording, Phrasing

The choice of words should support and not interfere with the interaction between human and machine. For this purpose, the wording of the system prompts has to be short and concise, putting the answer in the caller's mouth. By imitating the IVR system's wording, the user will answer briefly and concisely too, thus facilitating the job of the speech recognition engine. The designer must also take care that the options don't sound too similar and that the speech recognition time outs are sufficient long.

The results of the Wizard of Oz tests, performed by "Siemens AG, Corporate Technology, User Interface Design Center"([Hempel 2006a], [Hempel 2006b]), point out that good enunciation, i. e. a clear and concise way of speaking, is of particular importance for elderly people.

If a caller with a limited ability to interact with the IVR system has to be assumed (e. g. elderly people) or if the dialogue history shows signs of an impaired interaction between human and machine, the IVR system's prompting can be modified by:

- Raising the volume, lengthening the time-outs for speech recognition to more than 3 sec. (commonly 2–3 sec.) and reducing the speed of speech.
- Use of generally intelligible, self-explanatory expressions, a carefully use of technical terminology and emphasis of the current keywords.

When considering the fulfillment of expectations and the needs of a specific user group the following strategies are recommended:

- To adapt the speed of speech and the pauses for speech recognition to the speed of speech and the response times of the caller or, if no caller related data is available, to the average scores of the user group that the caller belongs to.
- To adapt pitch, intonation and vocabulary to the values that are generally accepted for that specific user group, e. g. using a rather polite, elaborated, high level and cooperative conversational style for females and a rather competitive, lean and direct conversational style with an emphasis on the imperative for males.

1.5.8 Persona Design

To ensure smooth interaction, the artificial agent has to be fast and direct if the interaction runs smoothly, and should become sympathetic and helpful if the interaction comes to a deadlock. It should be able to control the interaction without being arrogant or overbearing. It should neither apologize to the caller nor accuse them of anything. If problems persist, it should immediately hand over to a human agent. If possible, it should not only be prepared for a special range of tasks, questions or problems. It should also be prepared to handle unexpected events.

The caller infers a lot of information from the artificial agent's voice. "Social Behavior" comprehends all of the characteristics of an artificial agent which contribute towards its naturalness and personality:

- the artificial agent's age, gender and geographic background,
- their social function within the interaction (expert, assistant, teacher,....),
- quirks, foibles and emotions create personality – ultimately the artificial agent is allowed to show quirks, foibles and emotions, too.

To design an appropriate persona, it is not enough to divide customers into target groups with a tailored adaptation of the persona. The question about the context of the application is at least as important, that is the question as to why the caller entered the dialogue. As mentioned before, expectations and needs differ enormously depending on the scenario; for example on whether the customer has to deal with an infotainment system, or with a banking application. The design of the persona has to be particularly aware of this.

Callers create a mental image of the personality of the speaking artificial agent. Depending on the target group and the context of the application, the persona is able to use very different accents, beginning with "professional", "competent" or "reliable" up to "sympathetic", "trustworthy" or "helpful". On the basis of the results of the Wizard of Oz and usability tests, performed by "Siemens AG, Corporate Technology, User Interface Design Center" ([Hempel 2006a], [Hempel 2006b]), even more accurate conclusions can be drawn.

Age:

- Basically, middle-aged agents are recommended. The age of an agent should never be less than 20 years or more than 60 years.
- For entertainment applications rather younger agents can be selected.

Speaking style:

- Extremely casual slang language won't be accepted, even by teenagers. Particularly females are reacting adversely to this.

Gender:

- In general there are no distinct trends to be observed – female agents seem to be especially suited for applications in the context of information and entertainment and for children as a target group. For young males as a target group and for applications which have to do with the handling of problems, male agents are rather ranked positively.
- To ensure an authentic design of persona, male and female agents should actually use typically male or female conversational styles; a rather polite, elaborated, high-level and cooperative conversational style for female agents and a rather competitive, lean and direct conversational style with emphasis on the imperative for male agents.

Naturalness versus artificiality:

- The design of a persona has to reflect a balance between naturalness on the one hand and transparency and functionality on the other.
- Too much artificiality, e. g. caused by using a text-to-speech engine instead of pre-recorded, natural prompts, meets with disapproval, especially from females.

- Too much naturalness on the other hand raises expectations too high compared to the real possibilities of the human-machine interaction, especially for children and older people. It should be carefully explained, that the caller is not interacting with a human but with a machine, to avoid possible misunderstandings.

Social role:

- For applications handling problems (e. g. trouble shooting), the persona has to display competence.
- In general, if the artificial agent shows equitable, conversational behavior based on partnership, then it is considered acceptable.

1.5.9 Background Music, Jingles

Another tried and tested way of addressing a target group is the introduction of music, for example as background music, through jingles or to break up waiting loops.

A comprehensive analysis of consumers by Bauer Media KG in 2006, initiated by the MIZ (Deutsches Musikinformationszentrum des Deutschen Musikrats [http://www.miz.org/statistiken.html]), provides information about music preferences in Germany, broken down by age and gender. More information about the use of music can also be gathered from studies by "Siemens AG, Corporate Technology, User Interface Design Center", based on a Wizard of Oz test and a usability test of a prototype speech application, implemented by T-Systems ([Hempel 2006a], [Hempel 2006b]).

The results show very plainly that music can polarize more than any other criterion. To meet the target group's expectations, music has therefore to be selected very carefully. The basic results at a glance:

- Regarding age, people younger than 30 years and older than 60 years build two comparatively homogeneous groups, in complete opposition to one another.
- For people in Germany below 30, music genres like pop and rock and, with restrictions, hip hop, techno, trance, dance floor, hard rock or heavy metal are popular. For people older than 60, genres like "Schlager", "Volksmusik", oldies, evergreens, classical music, opera, operetta or country are popular.
- Within the group of people between 30 and 60, a crossover between the opposing profiles takes place. The preferences in this age group are therefore very heterogeneous.

- Males in Germany prefer genres like rock, hard rock, heavy metal or country. On the other hand, females in Germany prefer genres like "Volksmusik", "Schlager", oldies, evergreens, musicals, opera or operetta.

Of course, this data applies within the cultural context of Germany and will differ from one country to another. However, some basic tendencies can be transformed into some basic recommendations for the use of music in speech applications.

- The context of the application is again important, thus the question as to why a caller enters the dialogue, plays an important role when selecting the music. As stated above, expectations differ if the customer is dealing with an infotainment system or with a banking application.
- In the case of a broad, heterogeneous clientele, the selection of neutral music that is not polarizing is recommended (e. g. pop music with a soft touch, classic music with a popular coloring).
- For the selection of music, particularly the age of the user must be considered. The gender seems to be of minor importance.

1.5.10 Content

Knowing more about the caller, e. g. the age and gender of the caller, thanks to information that can be obtained from a customer database or from classification of the speaker, plays a significant role when defining the content, teasers, advertisements or marketing strategies that a speech application should offer. Determining the user's preferences and the correlation of the preferences with socio-demographic parameters, such as age or gender, may help to optimize the content and the marketing, which can be tailored to the target group's preferences, expectations and needs.

1.5.11 Multilingual Services

Up to now, speech applications have offered monolingual services, thus excluding all of those callers who are not familiar with that language. In view of globalization and an increasing multicultural population it is no longer acceptable to exclude a significant part of the potential clientele. Offering multilingual services thus represents an increasing market. Seen in this perspective, the language identification technology is the door-opener for catching foreign language speakers before they hang up. The solution is to automatically detect the caller's language and switch the language of the IVR system or dispatch the caller to an agent who speaks their native language.

Within a usability test, performed by "Siemens AG, Corporate Technology, User Interface Design Center", first experience with multilingualism in IVR systems was gathered. The test mentioned was based on a multilingual prototype, implemented by T-Systems. Overall 20 German, 18 English and 16 Turkish test persons participated.

The prototype was provided with automatic language identification. Instead of asking the caller which language they would like to hold the conversation in, language identification allows the caller to answer directly in their native language. The identification of the caller's language then initiates the switching of the language of the IVR system.

The focus of the test was to address the question of how to design the IVR system's welcome, so that the caller is immediately aware that they can answer in their native language. Different versions were implemented and tested.

The tests provided some important results which have to be considered when designing a multilingual IVR system.

- The invitation to use your native language has to be presented directly and immediately to avoid the caller hanging up.
- The invitation to use your native language must be presented in the language concerned, to be sure that the caller really understands.
- The invitation to use your native language must be spoken by native speakers with no accent, to get the required acceptance.
- The caller may be worried about extra costs, thinking of an international call when speaking in another language. So it might be helpful to tell the caller that no extra costs arise from the call.

1.6 Discussion and Conclusion

Quality is the fundamental requirement for the commercial success of IVR systems. To prevent callers from breaking off the communication and hanging up, the IVR system's overall quality has to meet the customer's expectations. The basic condition is a working, error-free and fully functional IVR system. Stability and performance has to be ensured for all components of the IVR system. Another important factor is whether the application will be accepted by the customer. The essential question here is whether the customer will call again. To maximize acceptance, the IVR has to be aware of the customer's abilities, needs and expectations.

As we have seen, the development of speech applications is a highly complex process. To ensure that the IVR system's overall quality meets the customer's expectations, each step of the development process should be provided with its own specific quality tests. Note that costs for quality assurance, testing and tuning will take up at least 50% of the total budget. Often underestimated in the past, the calculation of these costs plays a fundamental role, if you wish to avoid low quality in the IVR system or a budget overrun.

To estimate the caller's ability to interact with the IVR system, very different parameters can be considered, depending on their availability. If a low frequency of usage is to be expected, the vast majority of the users will be new and not familiar with the IVR system. Other parameters such as the caller's age, emotion and speech (native language, accent, dialect, sociolect), the frequency of certain dialogue events (e. g. interruption, no match, no input or help) or the environmental conditions might also contribute to an estimate of the caller's ability to interact with the IVR system. To ensure that the caller can cope and to reflect the caller's ability to interact, we recommend changes to the dialogue which might affect the strategy for the dialogue (mixed initiative, directed dialogue, confirmation and feedback), the menu tree, the system output (speech volume, speed and pauses, emphasis of keywords) or fall-back strategies (touch-tone, assistance from a human operator).

If the objective is to fulfill the callers' expectations and needs, the application context and a socio-demographic characterization of the caller have to be at the center of attention. Expectations and needs differ if the customer has to deal with an infotainment system or if entering a safety critical and security sensitive business process such as a banking application. For a socio-demographic characterization we have to consider the caller's age, gender or membership of a particular socio-cultural milieu. Changes to the dialogue might affect feedback and confirmation strategies, the order and hierarchy in which the themes and contents are presented, the design of the system's persona and voice as well as the selection of appropriate background music, teasers, jingles or advertisements.

Native language, accent, dialect, sociolect and speed of speech contribute to the linguistic characterization of the caller. This might cause the system's language to change or a transfer of the caller to an agent who speaks the caller's language.

In the above section we also described how much information we can get about the user, even when there is no user profile database available. Speaker classification is the key technology here. Whereas gender and language identification seems to be mature technologies, age and anger recognition still shows significant error rates and the identification of accents, dialects or sociolects is just starting. Nevertheless further research will lead to an increase in performance in future, thus helping to close this gap in the dialogue designer's knowledge.

Acknowledgements

The work described in this section was funded by Deutsche Telekom AG, Laboratories, represented by Dr. Udo Bub, through a series of research contracts awarded to the Advanced Voice Solutions group at T-Systems Enterprise Services. I would like to acknowledge the support of Katja Henke, Dr. Roman Englert and Dr. Florian Metze of the project field "Intuitive Usability" at T-Labs. I would like to express my thanks to Dr. Thomas Hempel from the Siemens AG, Corporate Technology, Competence Center "User Interface Design" for the fruitful co-operation with Deutsche Telekom AG. I would also like to thank Joachim Stegmann and all of my colleagues from the advanced voice solution team at T-Systems Enterprise Services for their collaboration and useful discussions during the last years. The studies on interindividual differences were supported by Markus van Ballegooy from T-Mobile Germany and Wiebke Johannsen. I acknowledge my niece Teresa Monkkonen from Rice University, Houston TX, my wife Federica Biagini and Kim Booth for revising, as well as Joachim Stegmann and Florian Metze for final approval of this work.

References

Anastasi, A. (1976). Differentielle Psychologie. Vol. II, Beltz, Weinheim, 1976.

Asendorpf, J.B. (2003). Person/situation (environment) assessment. In R. Fernández-Ballesteros (Ed.), Encyclopedia of Psychological Assessment. Vol. 2, London, U.K., Sage, pp. 695–698.

Baltes, P. B. (1990). Entwicklungspsychologie der Lebensspanne: Theoretische Leitsätze. Psychologische Rundschau, 41, 1990, pp. 1–24.

Bickmore, T.; Cassell, J. (2005). Social Dialogue with Embodied Conversational Agents. In J. van Kuppevelt, L. Dybkjaer, & N. Bernsen (Eds.), Advances in Natural, Multimodal Dialogue Systems, Springer Netherlands.

Braun, F. (2004): Reden Frauen anders? Entwicklungen und Positionen in der linguistischen Geschlechterforschung. In K. Eichhoff-Cyrus (Ed.), Adam, Eva und die Sprache, Beiträge zur Geschlechterforschung. Mannheim, Dudenverlag, pp. 9–26.

Buisine, S.; Abrilian, S.; Martin, J-C. (2004). Evaluation of multimodal behaviour of embodied agents. In Z. Ruttkay and C. Pelachaud (Ed.), From Brows till Trust: Evaluating Embodied Conversational Agents. Kluwer.

Burkhardt, F.; Ajmera, J.; Englert, R.; Burleson, W.; Stegmann, J. (2006). Detecting anger in automated voice portal dialogues. Proc. Interspeech 2006, ISCA, Pittsburgh, PA, USA.

Burkhardt, F.; van Ballegooy, M.; Englert, R.; Huber, R. (2005). An emotion-aware voice portal. Proc. 16. Conference for Electronic Speech Signal Processing (ESSP) 2005, Prague, Czech Republic.

Canada, K.; Brusca, F. (1991). The technological gender gap: Evidence and recommendations for educators and computer-based instruction designers. Educational Technology Research & Development, vol. 39, no. 2, pp. 43–51.

Catrambone, R.; Stasko, J.; Xiao, J. (2004). ECA as user interface paradigm. In Z. Ruttkay and C. Pelachaud (Ed.), From Brows till Trust: Evaluating Embodied Conversational Agents, Kluwer.

Cerrato, L.; Falcone, M.; Paoloni, A. (2000). Subjective age estimation of telephonic voices. Speech Communication, vol. 31, no. 2–3, pp. 107–102.

Duda, R. O.; Hart, P. E.; Stork, D. G. (2000). Pattern Classification. 2nd ed., Wiley Interscience.

Fraser, J.; Gibret, G. (1991). Simulating speech systems. Computer, Speech, and Language 5, pp.81–99.

Gilly, M. C.; Zeithaml, V. A. (1985). The elderly consumer and adaptation of technologies. Journal of Consumer Research, vol. 12, pp. 353–357.

Gomez, L. M.; Egan, D. E.; Bowers, C. (1986). Learning to use a text editor: some learner characteristics that predict success. Human- Computer Interaction, vol. 2, pp. 1–23.

Günthner, Susanne (1997). Zur kommunikativen Konstruktion von Geschlechterdifferenzen im Gespräch. In Braun, F. /Pasero, U. (Eds.), Kommunikation von Geschlecht – Communication of Gender. Pfaffenweiler, Centaurus, pp. 122–146.

Hempel, T. (2006a). Usability of Telephone-Based Speech Dialogue Systems as Experienced by User Groups of Different Age and Background. In: 2nd ISCA/DEGA Tutorial and Research Workshop on Perceptual Quality of Systems, Sept. 04th–06th, Berlin, Germany, International Speech Communication Association: Bonn, Germany, pp. 76–78.

Hempel, T. (2006b). Umgang von mittelalten und älteren Nutzern mit telefonbasierten Sprachdialoguesystemen. In: Usability Professionals 06/Mensch & Computer 2006 – Mensch und Computer im Strukturwandel, Sept. 3rd–6th 2006, Gelsenkirchen, Germany, University of Applied Sciences.

Kienast, M.; Paeschke, A.; Sendlmeier, W. F. (1999). Articulatory reduction in emotional speech. Proceedings Eurospeech 99, Budapest, pp. 117–120.

Krämer, N. C.; Rüggenberg, S.; Meyer zu Kniendorf, C.; Bente, G. (2002). Schnittstelle für alle? Möglichkeiten zur Anpassung anthropomorpher Interface Agenten an verschiedene Nutzergruppen. In M. Herzceg, W. Prinz & H. Oberquelle (Ed.), Mensch und Computer 2002, Teubner, Stuttgart, pp. 125–134.

Krämer, N.C. (2001). Bewegende Bewegung. Sozio-emotionale Wirkungen nonverbalen Verhaltens und deren experimentelle Untersuchung mittels Computeranimation. Lengerich, Pabst.

Lee, C.M.; Narayanan, S. (2005). Towards detecting emotions in spoken dialogs. IEEE Transactions on Speech and Audio Processing, 13(2), pp. 293–302.

Levin, T.; Gordon, C. (1989). Effects of gender and computer experience on attitudes toward computers. Journal of Computing Research, 5(1), pp. 69–88.

McBreen H. (2002). Embodied conversational agents in e-commerce. In Socially Intelligent Agents: Creating Relationships with Computers and Robots. Kluwer Academic Publishers.

Metze, F.; Ajmera, J.; Englert, R.; Bub, U.; Burkhardt, F.; Stegmann, J.; Müller, C.; Huber, R.; Andrassy, B.; Bauer, J. G.; Littel, B. (2007). Comparison of four approaches to age and gender recognition for telephone applications. Proc. ICASSP 2007, IEEE, Honolulu, Hawaii.

Mulac, A. (1999). Perceptions of women and men based on their linguistic behavior: The Gender-Linked Language Effect. In Pasero, U. /Braun, F. (Eds.), Perceiving and performing gender. Opladen, pp. 88–104.

Paterno, F.; Mancini, C.; Meniconi, S. (1997). ConcurTaskTrees: A diagrammatic notation for specifying task models. Proceedings Interact'97, Chapman&Hall, July, Sydney, pp. 362–369.

Rabiner, L. R. (1989). A tutorial on hidden markov models and selected applications in speech recognition. Proceedings of the IEEE, vol. 77, no. 2, February, pp. 257–286.

Reynolds, D. A.; Campbell, J. P.; Campbell, W. M.; Dunn, R. B.; Gleason, T. P.; Jones, D. A.; Quatieri, T. F.; Quillen, C. B.; Sturim, D. E.; Torres-Carrasquillo, P. A. (2003). Beyond Cepstra: Exploiting High-Level Information in Speaker Recognition. Proc. Workshop on Multimodal User Authentication in Santa Barbara, California, pp. 223–229.

Rudinger, G. (1994). Ältere Menschen und Technik. In Kastner M. (Ed.), Personalpflege: Der gesunde Mitarbeiter in einer gesunden Organisation. Quintessenz, München, pp. 187–194.

Schölkopf, B.; Smola, A. (2002). Learning with Kernels: Support Vector Machines, Regularization, Optimization and Beyond. MIT Press, Cambridge, MA, USA.

Sproull, L.; Subramani, M.; Kiesler, S.; Walker, J. H.; Waters, K. (1996). When the interface is a face. Human-Computer Interaction, vol. 11, pp. 97–124.

Sproull, L. S.; Kiesler, S.; Zubrow, D. (1984). Encountering an Alien Culture, Journal of Social Issues, 40(3), pp. 31–48.

Strong, E. K. Jr. (1943). Vocational interests of men and women. Stanford University Press, Stanford.

SWR (2004). Media-Analyse 2004/II. Media Perspektiven, SWR.

Walker, M.; Langkilde-Geary, I.; Wright, H.; Wright, J.; Gorin, A. (2002). Automatically training a problematic dialogue predictor for a spoken dialogue system. Journal of Artificial Intelligence Research 16, pp. 293–319.

Chapter 2

After discussing challenges in the field of phone-based IVR systems, Jan takes us to another level: developing and using a speech user interface in the context of a 'real' living room environment. On the technical side this means that the acoustical environment is much harder to control than for a phone-based system. But a factor of equal importance is that users who use speech input may receive system response by both audio and visual means. Indeed, speech input may be only one form of interaction in a 'real' environment. Finally, spatial and semantic issues are also likely to play a vital role.

Dipl.-Ing. Jan Krebber

Jan Krebber is an electrical engineer working on issues in the field of communication acoustics. His research focus is combining voice control and usability. Jan carried out most of his studies at the Ruhr-University, Bochum (Germany) and Dresden Technical University (Germany). Jan now is with Sysopendigia, Finland.

Rosa Pegam, M.A.

Rosa Pegam studied computational linguistics and communication focusing on man-machine interaction. For several years she worked on studies on voice-control and usability at Bochum Ruhr-University. Here, her main research focus is managing user expectations in spoken-dialogue systems. She is currently working in the field of Internet Technologies as a concepter and web designer.

2 Experiences of Designing a Speech User Interface for a Smart Home Environment

Jan Krebber
Ruhr-Universität Bochum, Germany, now with Sysopendigia, Helsinki Finland

Rosa Pegam
Ruhr-Universität Bochum, Germany

2.1 Introduction

The EC-funded project INSPIRE took place during years 2001–2004 and aimed at building, assessing and evaluating a smart home speech dialog system. The two most important reasons for building the system were to study new speech dialog system building techniques and to establish a new taxonomy for telephone based speech dialog systems. After the project was finished the system continued as a core system for further research-like user modeling approaches. This chapter will explain some fundamental aspects for designing a smart home speech dialog system in general and it shows the development process of INSPIRE in the end of each section. The chapter discusses basic requirements, dialog strategies, technical realization, final check, basic advices and closes with a glance of all points discussed.

The present work has been performed in the frame of the EC-funded IST-project INSPIRE (IST-2001-32746). Partners of INSPIRE are: Knowledge S.A., Patras, Greece; WCL, University of Patras, Greece; IKA, Ruhr-University Bochum, Germany; ABS Jena, Germany; TNO Human Factors, Soesterberg, The Netherlands; Philips Electronics Nederland B.V., Eindhoven, The Netherlands and EPFL, Lausanne, Switzerland. The authors would like to thank all INSPIRE partners for their support in the experiments and for fruitful discussions.

2.2 Basic Requirements

2.2.1 Aim of the System – Cornerstones

When the initial plans and aim of the smart home speech dialog system is clear the first concrete approach could be to sketch 5–20 different scenarios depending on the complexity of the system. From these typical scenarios one may clarify the expected cornerstones which are related to:

- Expected user groups,
- Devices to be controlled or tasks to be fulfilled,
- Communication interfaces and modalities,
- Time and financial limits,
- Commercial or research system.

The first check could be done by reflecting the first question: "What is the aim of the system?" and ensuring that the question is answered by every cornerstone.

In case of INSPIRE the aim was to build a smart home system with a unique user interface for "home" and "remote access" control for evaluating different questions about usability of smart home speech dialog systems. Some certain assumptions were made:

- Expected user groups
  - The users are expected to be novices who do have little or no experience with speech dialog systems.
  - The system should support the technically inclined user by simplifying the control of electronic devices.
  - The system is of assistive or supportive technology, to assist elderly or handicapped people, or to support the user in controlling a complex environment or a complex device, and finally in merging the control of different devices into one unique user interface.
  - There is only one user at the time allowed to operate the system.
  - External users like visitors or intruders are not allowed to use the system.
- Devices to be controlled or tasks to be fulfilled
  - Devices of different complexity and with immediate response to reduce experimental time, e. g. controlling a fan because of immediate feedback instead of an air conditioning system.
  - Approximately five different types of devices.
  - Single task oriented system.

- Communication interfaces and modalities
  - Use the latest speech technology and to reduce complexity, barge-in is not excluded.
  - Control via microphone arrays inside the house.
  - Control via remote access by telephone.
  - Output via direct response of the device or by speech.
  - Output via telephone handset for remote control.
  - Output via TV screen could be an option for in-house usage.
- Time and financial limits
  - Time schedule according to the project plan.
  - Funding mainly covers labor and allows only restricted amount of technical realization.
- Commercial or research system
  - Research system to examine the usability and to contribute to build taxonomy for speech dialog systems.
  - After further modifications, the system may go into field studies.

In the following we will discuss the design aspects of a smart home system with speech user interface. There is an example discussion at the end of each subsection to show how the decisions were taken based on the cornerstones in the INSPIRE project.

2.2.2 Expected User Groups

To design the speech dialog system it is expedient to create a set of typical user identities or at least profiles for further examination. To describe a typical user the biography should cover:

- Gender,
- Age,
- Regional provenance,
- Job position,
- Financial status,
- Health status,
- Family status,
- Bigger plans for the future,
- Leisure activities,
- Additional features as long as they may concern the later usage of the speech dialog system.

These descriptions of typical users are necessary for later example dialogs and for building the flow charts. One of the most underestimated points is the age of the user (Docampo Rama 2001). The most relevant breakpoint in lifetime could be the transition from work to retirement. Unfortunately there are only very few reports about usability or quality experiments with elderly as a target group. In the context of elderly users some of the often forgotten features when building the description of the typical users are:

- Vision or hearing impaired users,
- Physically handicapped users,
- Users with different experiences; novices or experienced, newbies or geeks,
- Expected group of the adoption process; innovators, early adaptors, early majority, late majority or laggards (Möller 2000).

After generating the description of the typical user the next step is to create typical situations or scenarios for using the speech dialog system. In case of INSPIRE these were related to using the system via remote access from the office, car, or directly at home. These three different environments show very different challenges for one task, e. g. calling in a car does not allow any keyboard handling at all.

The expected user group for the INSPIRE system was restricted to normal hearing and normal sighted users, but the concept explicitly included physically challenged users. The system was expected to be used by novices as well as by experienced users.

One of the prototypical users in the INSPIRE project is Michael Hoffmann. He is 32 years old and living with his wife Anna in North-Rhine Westphalia, Germany. He is working as a programmer in a small company and earning around 45000 Euros per year. Michael and Anna have no children so far, but they plan 1–2 children in the near future. Michael enjoys good health, he loves reading the newspaper and all kinds of books.

2.2.3 Corporate Design

It should be clarified whether there are any corporate design issues to attend to. The corporate design can affect to the persona of the system, the audio design or to specific keywords to mention some points. In case a celebrity or a certain actor is doing commercials for the client, he or she could represent the persona and finally give the voice to the system. Also commercials as such or websites (e. g. the wording used in the website) may be essential for later keywords and dialog structures.

For the INSPIRE system there was no corporate design as it was a purely academic system without any interest in implementing corporate design aspects into the speech dialog system. Besides, the web-space was more for the overall project related issues than for the system as such. The only corporate design issue was the naming of the avatar or the ghost; his name was Inspire.

2.2.4 Persona of the System

The persona describes the intended nature or the character of the system. It represents the system and finally the company behind. Again, the persona of the system should have a biography from which several later design aspects can be drawn. In addition, the metaphor of the persona may play a role. There are mainly three different possibilities to be considered:

Intelligent Devices
Each device is represented by a certain persona or at least each device gets its own loudspeaker to attract the attention of the user to the chosen device.

Ghost
All devices are operated by an invisible servant.

Avatar
All devices are operated by a visible servant, which requires several screens in every room, so that the avatar can always be seen from any position.

Tests have shown that the avatar requires much technical effort, especially for lip-and-speech synchronization and for optimizing the availability of eye contact at nearly every position in the house. Moreover, the users felt observed by the always present avatar.

The "intelligent devices" metaphor shows that the drawback is the non-uniform handling of the system, since there may be a "master-device" which hands over the dialog to the chosen device after it has been indisputably selected. Furthermore, every single device needs its own loudspeaker to make the user to believe that the device is "alive".

In later INSPIRE versions only the "ghost" metaphor was finally implemented mainly because of the drawbacks shown by the other metaphors. It was developed as a servant system with a servant attitude resulting in a polite wording.

2.2.5 System Output

The system output is one of the most considerable challenges in speech dialog design. On the one hand, there is all the information the system should confirm, questions about still missing information and the presentation of final actions or solutions. On the other hand, there is the wish to keep the dialog as clear and short as possible.

So far different kinds of short acoustical or visual "embedded signs of understanding"[1] are out of reach for the system developer until today because of two reasons: The system still suffers from delay times caused by the automatic speech recognizer (ASR). Secondly, the ASR is lacking the feature of anticipation. During a conversation a human partner can indicate by visual or tonal cues that he is familiar with the topic and anticipate what the conversation partner will say next. He can verify a certain task right in a moment when the word is spoken by nodding or simply saying "okay" or "yes". Compared to that speech dialog systems need approximately 2–3 seconds for ASR to detect the end of utterance and for dialog manager to process the recognized speech signal. However, the delay would result in a sign of understanding at an unexpected part of the interaction between user and speech dialog system or may just look unsuitable or sound irritating. It is a kind of art to find the optimum trade-off which will mostly be a compromise between provided information and displeasure.

It may be worthwhile to hire a professional "listening text" writer. One should be aware that a "listening text" differs quite much from normal prose. Therefore, the existing persona of the system and the expected user groups will pay off, as the writer can work on that persona, and he or she knows how to formulate certain phrases for the expected user group.

In case of the INSPIRE system the prompts were reduced to a minimum information and cut to shortest phrases possible. Even the system lost some of its politeness, the decrease in annoyance of the users was remarkable and advantages offset the disadvantages. This can be shown by the task follow-up phrase, which was always given after a successful task, asking for the next task by "What else can I do for you?"[2] This nice and polite prompt as used in the first version was shortened to a more colloquial style prompt "Anything else?"[3] in the following versions.

---

[1] "Embedded sign of understanding" or "back channel item" is any kind of feedback of speaker A during the utterance of speaker B. Speaker A wants to show to speaker B that he understood and speaker A wants to keep the dialog short.

[2] German: "Kann ich noch etwas für Sie tun?"

[3] German: "Noch etwas?"

2.3   Dialog Strategies

2.3.1   Keywords

One of the basic tasks in creating dialog strategies is to identify all the relevant keywords. It is worth paying attention to special wordings according to the expected user groups. The corpus should contain all the relevant keywords, keyword combinations and phrases related to each task. The analysis of the corpus can point out certain communicative situations which need extra caretaking.

One example of these special communicative situations was the German word "Nachrichten", which has two meanings: news or messages. News is related to the TV whereas messages are related to the answering machine. This homonym is one of the communicative situations which need extra caretaking. This was solved by associating additional keywords like "watch the news"[4] and "listen to the messages".[5]

2.3.2   Information Gathering Strategy

There are some strategies how to collect all the necessary information for fulfilling a system driven task. As shown in Fig. 2.1 one basic strategy of speech dialog systems is often based on attribute-value pairs (Rajman 2004). Information gathering may vary for each system and ranges from asking for each attribute's value in a specific order, to filling the attributes in a completely unstructured way. Something between those two extremes is normally taken for a task driven "conversational" speech dialog systems.

There might be differences in the "cooperation" of the system, like in supporting the input with help or how to provide alternative solutions in case of mismatch, but also in quitting a dialog because of security considerations.

The INSPIRE system was designed as a task oriented and cooperative system allowing the user an unlimited initial operation. This initial operation was mainly intended to specify the devices and to collect as much additional control information as possible. Therefore the INSPIRE system was based on Generic Dialog Nodes (GDN). Each GDN can be filled with exact one attribute-value pair. In case of the example as presented in Fig. 2.1 the first GDN is filled with Device=fan and the second GDN gets the attribute-value pair Action#1=switch_on whereas all other attributes stay empty.

---

4   German: "Nachrichten sehen"

5   German: "Nachrichten abhören"

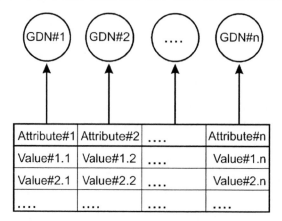

Attribute	Device	Action#1
Value	fan	switch_on

Fig. 2.1. Generic dialog nodes (GDN) as associated with the attribute-value pairs according to Rajman (2004) and attribute-value matrix sample

In case of incoherencies, like specifying a certain device with an action which could not be done by the device (e.g. Device=fan AND Action#1=mute_loudspeaker), the system proposed several solutions referred as *incoherency strategy*. This could be either changing the device or changing the action. The three possible solutions are shown in Table 2.1.

Table 2.1. Example of possible solutions to solve incoherencies

Solution	Device	Action#1
1	fan	switch_on
2	fan	switch_off
3	TV	mute_loudspeaker

The system was intended to keep the dialog alive as long as possible and not to quit the dialog because of incoherencies. In case there were too many incoherencies in one utterance, the system chose the three most likely e.g. the most performed actions as proposed solutions. Those three solutions were offered to the user as a list from which he could choose the final task action.

2.3.3 Standard Dialog Handling

The following dialog strategies should be taken into account when developing a speech dialog system.

Assertion
The system assumes that it has understood and informs the user which values were filled in (e. g. in INSPIRE "I understood fan as device").

Repetition
The system asks the user to repeat his utterance because no part of the utterance matched with anything in the keyword database,[6] or the system reacts to the request for repeating its last prompt.

Help request
The system reacts to a help request. Help can either be triggered explicitly by the user (e. g. in INSPIRE "Which lamps can I operate?") or be provoked implicitly because the user made at least two incomprehensible utterances in succession. This latter case can occur in certain situations, for example while using the INSPIRE system the user did not twice specify the location of a specific lamp, which the system in turn interprets that it should support the user by listing the possible options to progress in the current situation: "The following lamps can be operated: all lamps, the table lamp, the white or the yellow floor lamp."

No input
The system does not receive any input either because the user speaks in a too quiet voice or because the user does not utter anything at all during a predefined time span.[7] In such cases the dialog manager has to assume that the user feels insecure or is not confident what to say and consequently react by saying e. g.: "I could not hear you" and subsequently provide help to the user.

System non-understanding and out of context
The system could not match any of the user inputs and hence informs the user about its misunderstanding for example by saying: "I could not understand you". This is followed by a help prompt providing the user with all the options he has at the current point of interaction.

---

[6] INSPIRE was not working with a dictionary or single keywords, but rather with phrases of keywords.

[7] In INSPIRE this span was 20 seconds although other dialog systems may require different periods of time until *no input* handling.

Dialog dead end management strategy
In case no solution for the requested task can be found the system provides alternatives within the current context and asks the user to choose a suitable one.

Confirmation strategy
User verifies the values for irreversible actions.

Dialog termination strategy
The system decides when a solution can be proposed and therefore terminates the dialog. This strategy is needed in case there is only one possible attribute-value pair left because of certain system states.

Incoherency strategy
The system received contradictory values and enters a clarification dialog as already explained above.

Canceling
User requests canceling or restarting the dialog.

2.3.4 Exceptional Dialog Handling

Exceptional dialog handling is required in case the system comes up against limiting factors. The limitation of a speech dialog system is mainly related to its non-existing knowledge of the world and to the fact that the system is usually not capable of solving more than one task at a time. The latter point is necessary to ensure the system stability and a clear dialog structure. Differences in the system limitations can be made between dynamic system limitations which have linguistic origin, and technical system limitations which are referred as static system limitations.

2.3.4.1 Dynamic System Limitations (Linguistic)

References Related to Static Objects
From linguistics point of view the most challenging question is related to different kinds of references. As mentioned before, the system is not capable of resolving references in the same way as humans can. When the system is examined closely, it is not able to resolve any references at all since in general the resolution of references is made possible by knowledge of the world, which humans have established by making experiences.

There are several mechanisms for resolving anaphoric references through pronouns in computational linguistics.[8] For spoken language dialogs with human-machine interaction those mechanisms have to be adapted and altered according to differences which exist between spoken and written language (e. g. changing of the grammar structures and covering of prosodic effects). So far only very few dialog systems actually make use of reference resolution because for the last few years dialog systems were mainly used for call-routing or information retrieval.

For systems like INSPIRE that were designed to master more complex task, reference resolution would be of notable use. The interpretation of temporal, thematic and spatial context reduces confusion since ambiguities will be handled and users will avoid repeating themselves as often because of clarifying the dialog flow. The following extract as an example of a dialog in INSPIRE in Table 2.2 illustrates references which point to items across several turns.

Table 2.2. Sample dialog referring to lamps

S1:	What else can I do for you?
U1:	Switch on the lamps.
S2:	I understood lamp as device and switch on as task. Which lamp do you want to operate?
U2:	Left.
S3:	switches lamp on; Anything else?
U3:	Reduce the brightness of the lamp.
S4:	I understood lamp as device and down as task. Which lamp would you like to operate?
U4:	Left again.
S5:	dims lamp; What else can I do for you?

S system, *U* user

In this sample the user is first supposed to switch on a lamp and adjust its brightness afterwards. The user switches on the left floor lamp by specifying its location ("left") (U2). After this is done he refers to the same lamp by saying "reduce the brightness of the lamp" (U3), but does not specify the exact location. The system, however, has started a new dialog

---

[8] Gieselmann (2004) lists various approaches which have been developed between 1977 and 2003. Most of them work with rules which are based on pronouns and their antecedents by means of which it is possible to resolve more than 90% of all references.

and therefore deleted the history. Hence it cannot resolve which lamp the user addresses, although for a human this would be no problem at all.

In the INSPIRE system the history feature was mainly discarded to keep clear dialog structures. Even no temporal or thematic reference was possible there were some spatial references, e. g. referring the position of the lamp to the "left" of the sofa, or the lamp "on the table".

References to Dynamic Objects

Dynamic objects are objects going through frequent changes such as movie titles in a TV database. It can happen that users refer to the program they want to choose with a proper name (the actual title) when operating the TV or video cassette recorder (VCR). Instead of stating the desired parameters (e. g. day and broadcast time), they might say "I want to see Tagesschau" or "Record Star Trek", while the system will ignore these indications. Humans, on the contrary, can resolve titles easily. Every adult German,[9] for instance, is certainly aware of the fact that the news program "Tagesschau" is shown every day at 8 p.m. on channel ARD. Regarding "Star Trek" it gets a bit more difficult, however still the program is known well enough for the user to be able to tell that the interaction partner is talking about a movie. Consequently, at least the feature "movie" could be extracted by the system by only being told the title of the program. The machine does not have any world knowledge as humans do which is sometimes referred as *pool of experience*.

Homonyms and Ambiguities
Homonyms are a difficult task for a system developer. If interpreted in the wrong task context, the system will fail to interpret the keyword correctly thereby producing a mismatch and a breakdown of the dialog flow. Prominent examples from the INSPIRE experiments were the words "Nachrichten" and "heute". The word "Nachrichten" means both news and messages. Therefore, depending on the context, this keyword can trigger the task focus to the answering machine as well as to the TV. Consequently it was frequently the case that the system entered a task focus which was different to that expected by the user. The word "heute" refers to "today" as well as to the news program "heute" which can result in a similar confusion.

Ambiguous indications evoke confusion as in this example of a user of the INSPIRE system: "I would like to operate the lamp next to the TV set". Expressions of this kind confuse the system, since the user mentions two keywords referring to devices: the lamp and the TV set. It was impossible for the system to tell that "next to the TV set" was merely an indication of the location of the lamp and did not refer to the TV set as such. In this case

---

[9]   At the time the system was designed.

the natural language understanding module failed and one of the input words was either matched to a different context or completely ignored. The system therefore lacked the ability to interpret syntactic and semantic context. For humans, on the other hand, hardly any ambiguities exist. The meaning of a word can easily be determined by means of context interpretation. Finally, the problem was solved by forcing the system to process long chunks of utterances first and to check whether there were any indications related to other device positions, like "lamp next to the TV" in the keyword list was giving Device=Lamp AND Position=Table_Lamp, instead of Device=Lamp OR TV.

Synonyms
Synonyms are another issue which occasionally can evoke dialog flow breakdowns often due to confusion on the user side. Some users have explicitly stated that they expect the system to use the same wordings as they used (as usually happens in interactions between humans). The system, however, lacks the desired flexibility to adjust to the need of the user in this respect as shown in the INSPIRE sample dialog given in Table 2.3.

Table 2.3. Sample dialog related to confusion about synonyms

U1:	Electronic program guide.
S1:	I understood program as task. On which day is or was the program broadcast?
U2:	Turn on help function.

S system, *U* user

This is a typical example of the system using words which differ from those of user. This results in bewilderment on the user side and causes a breakdown as a consequence of which the user initiates a help request (U2). As a matter of fact the system has processed the user utterance correctly. The user, however, does not comprehend because he expects the system to confirm his utterance using the same words as he does (he would have expected the system to say something like "I understood electronic program guide").

Complex Grammatical Constructions
Especially elderly people use grammatical constructions which the system may not be able to process correctly. An example of such is the participial construction "I want the fan turned on". This kind of phrasing was not supported by early versions of the system. The system interpretation is based on keyword concatenations such as "turn on fan". The same holds

for negations as in "not the fan". From a sentence like this the system would only extract the device name and ignore the negation.

2.3.4.2 Static System Limitations (Technical)

The second major category consists of technical kind of system limitations in contrast to linguistic or dialogic problems. Naturally, the problems to be outlined now do have consequences regarding the dialog flow, but the reasons are to be found in the system architecture.

Misrecognitions

Misrecognitions can be split into two sections. The first section includes all the misrecognitions brought about by the ASR. These can be caused by many problems like bad acoustics, insufficient speech signal pre-processing algorithms or a bad or badly trained ASR. The second section of mistakes involves insufficient amount of keywords or key phrases for the speech understanding section. This hints at an incomplete keyword list. The following extract from an INSPIRE dialog shown in Table 2.4 illustrates a typical case of misrecognitions by the ASR, which creates confusion.

Table 2.4. Sample dialog of ASR misrecognition, caused by incomplete keyword table

S1:	I understood down. Please state the device you would like to operate.
U1:	Flame.
S2:	I could not understand you. Your options are: the fan, the blinds, the lamps, the TV set or the answering machine.

S system, *U* user

The misrecognition occurred in U1 and was induced by the ASR which matched the user utterance wrongly to a word, which moreover is not a known keyword ("flame" instead of "fan"). As a consequence of the misrecognition the system initiates a repair turn. A dialog like this results in annoyance or even frustration if it occurs too often within one interaction. Though, special care should be taken for the most relevant keywords like "yes", "no", "help", "quit" and numbers. Those should come with word accuracy close to 100%, as this is sometimes not the case for the ASR when receiving it from the vendor.

Modality

Even though users are made aware of the fact that the system reacts only to speech as input modality, at times they still find themselves disappointed because the system does not react to physical pointing or vocal deictic expressions[10] (e. g. "this lamp here on my right side"). From the INSPIRE experiments it turned out, that a large share of participants stated that they would like to have an additional remote control at their disposal for "in-house" usage. It should be stated again, that it was in the specifications to build a unique voice interface for home and remote use.

Excessive Demand
Excessive demands describe the user requests for two or more tasks within one task. This is a problem most of the speech dialog systems cannot resolve in a satisfying way. One example is given in Table 2.5, which was a frequently occurring situation in INSPIRE.

Table 2.5. Sample dialog of excessive demands

S1:	Anything else?
U1:	Switch on the yellow lamp and dim it.
S2:	What should I do with the lamp?

S system, *U* user

The user tries to include two commands in one utterance: First he wants to switch on the yellow lamp and secondly he wishes to dim it. For a human it is a clear case that the lamp should be switched on before it is dimmed down, but for the system these are two different tasks which cannot be merged to a single one due to the single task paradigm of the system.

At many points users are probably not even aware of the fact that they are uttering more than one command in one input utterance or they cannot differentiate when they can give several keywords and when they cannot. Single task focused system cannot process user input which aims at fulfilling several tasks at once. Even though after a while users probably find out which inabilities the system has, they forget those when they are concentrated or excited, or too involved in the interaction process. This user behavior is to be attributed to a lack of awareness regarding the limited abilities of the system (e. g. restricted keyword recognition within a certain focus). To be able to operate several devices at a time is still a feature that many users strongly demand.

---

[10] The focus of sight is expected to be aligned with the focus of the acoustics of the speech utterance. From this the relative position of the device could be derived.

2.3.5 Additional Modalities

After starting with a pure speech dialog system several other input devices or solutions which can be related to further modalities might be added, e. g. the following ones:

Remote control handset is a customized hardware which comes with moderate expenses and is for controlling some or all of the functionalities e. g. of the smart home environment. Remote control handsets can be difficult to configure for every individual system and they offer only limited updating possibilities. Also benefits are restricted since usually there will be no additional control items compared to speech, and many times the remote controls are not immediately at hand.

Smart mobile phone and *PDA* are standard hardware with customized software. They remain cheap since they are usually easily available. In addition to that they allow bidirectional communication and are easy to update. Depending on the model, screen size and keyboard limitations handling and usability may easily become an issue. The usage of a mobile phone or a PDA may be restricted to certain models or operating systems. Hybrid solutions of ASR or full dialog manager implemented in a mobile communication device can be the future solution, but they require high computational load for a full speech dialog system. However, many mobile phones already offer a basic ASR and speech synthesizer functionality.

TV screen is often already available in households. This results in no additional expenses except integrating the screen into the dialog manager. Even the TV screen is just for output only purposes, it may substantially support the interaction especially in case of incoherencies.

Touch screen is usually not available and obtaining it will cause extra costs. It remains questionable if the touch screen will replace a TV screen, since there are fingerprints on the screen because of usage and the sensor foil may cause visual disturbances when watching moving pictures on the screen behind. The touch screen could serve as an additional input and output device for possibly all tasks, which means that touch screens should be available in all rooms of the smart home environment to compete with or to support the speech dialog system.

Pointer with accompanied sensors or other positioning devices can be integrated into a remote control handset. What comes at high price are the sensors for receiving the pointer signal, since those should be integrated "invisibly" into each device and in the optimum case cover the complete surface of the device.

Direction of speech can be derived by sophisticated microphone array solutions. In case microphone arrays are distributed all over the boundaries of the living areas, the position and the direction of the talker can be estimated from the microphone signals. Supposing that the direction of speech is the same as the direction of look and the direction of attendance, this could resolve the problem of references related to static objects.

Direction of facing can be derived by sophisticated camera solutions. Cameras will come with additional cost as they are of no use for a speech dialog system. However, in case of multimodality they could be used in case gestures or facial expressions are taken into account. However, many users feel observed when a camera is in the room.

In case of the INSPIRE system a TV was finally integrated to support the electronic program guide for giving detailed program information. Nevertheless, it was a compromise to improve the system acceptance by giving up the speech only guideline for "home use". Moreover, the INSPIRE system offered only one TV screen for the entire home allowing displayed messages only in the living room when sitting at the TV. However, the integrated TV screen improved the system acceptance as shown by heuristic experiments with the INSPIRE system.

2.4 Technical Realization

2.4.1 Components of the INSPIRE System

The INSPIRE system was designed in a modular way allowing the project partners to work independently on certain blocks or to replace blocks with the Wizard-of-Oz methodology (WoZ).[11] The final system structure is depicted in Fig. 2.2.

The microphone signals were sent to the beam forming block. From there one speech signal went to the noise reduction block. After that the signal was sent to the ASR and to the speaker verification/speaker identification block (SV/SI). Besides, the ASR and the SV/SI blocks were connected to the telephone input, e. g. to the signal coming from the telephone handset microphone at sender side. The ASR delivered a data stream to the speech understanding module. This module matched the data stream from the ASR with the possible key phrases, as given by the key phrase list. The key phrases associated at least with one possible attribute-value pair.

---

[11] The Wizard of Oz, as described in the book "The Wonderful Wizard of Oz" by Lyman Frank Baum (1900), is a ruler who impresses the subjects with "magic" tricks all of which are based on imagination and physics. The WoZ technique refers to the scene "Don't pay attention to the man behind the curtain." In speech dialog systems the WoZ replaces a module, which is not at its full functionality, e. g. at the time of a test. He brings the speech dialog system to live by acting as the replaced module should do.

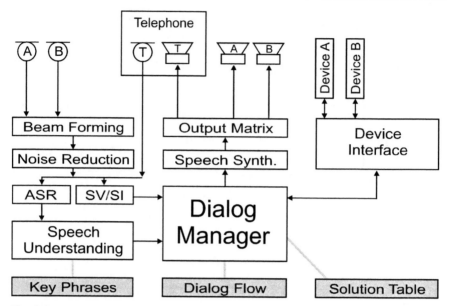

Fig. 2.2. Block diagram of the INSPIRE speech dialog system

Many times the key phrases linked to more than one attribute-value pair e. g.:

"I would like to listen to my new messages"[12] would have given the following attribute-value pairs:

- Device=answering_machine,
- Action#1=play_new_message.

Whereas "I would like to watch the news"[13] would have given:

- Device=TV,
- Action#1=switch_on,
- Action#2=select_channel,
- Genre=news.

The dialog manager is the core of a smart home speech dialog system. It puts the incoming attribute-value pairs into logical relation and decides when to perform which action. The action could be a speech output signal generated by the speech synthesizer and distributed via an output matrix to the loudspeakers. Alternatively, it could be a device action controlled via the device interface and the control bus.

---

[12] German: "Ich möchte meine neuen *Nachrichten abhören*"

[13] German: "Ich möchte die *Nachrichten sehen*"

The speech synthesizer of INSPIRE was based on concatenated speech which consisted of chunks of length between one word and a full phrase or sentence. The output matrix selected the loudspeakers according to the selected device, e. g. not every device had a loudspeaker, and nevertheless the sound was coming from the direction of the device. The device interface was built up in different versions. The first version was a simple USB-controller for controlling the devices by switching or dimming circuits, whereas the more sophisticated solution was a LON-bus[14] interface to control all devices via a local network.

The control tactics of the dialog manager were enumerated in the dialog flow list. This list contained all states which needed special attention of the system, e. g. at which point to provide help, or at which point in the dialog to offer several meaningful solutions. All possible solutions or final actions were listed in the solution table, e. g. the fan would have had two final states as shown in Table 2.6.

Table 2.6. Extract of the solution table showing all possible solutions for the device "fan"

Solution No.	Device	Action#1	Position	Action#2
105	…	…	…	…
111	fan	switch_on	not_available	not_available
112	fan	switch_off	not_available	not_available
121	…	…	…	…

2.4.2 System Extension

The extension of the system is linked to the question of system limitations which in turn can be caused e. g. by the vocabulary limitations of the ASR, the capability of the dialog manager, the amount of controllable sound output devices, the device interface capabilities or the additional controllers like a remote control handset. In case of a concatenated speech synthesizer the existing phrases may be another limitation for system expansion. Nevertheless, keeping the system in a modular way and grounding the control of the system in lists (key phrases, dialog flow, solution table), allow a system extension with reasonable amount of additional work. A challenge coming close to the system extension is the replacement of a module or device. Again, the modular structure pays off in this case, as only some parts of the system need additional caretaking and not the entire system.

---

[14] Local Operating Network, EN14908.

2.4.3 Performance

The quality of single component can only be tested on single-block tests, e. g. the component is tested as such before it is integrated into the entire system. When choosing the dialog manager itself, e. g. Rapid dialog prototyping methodology by Ecole Polytechnique Fédérale de Lausanne or CSLU Toolkit, one should be aware of the fact that those are made for small to medium size projects and will come soon to their limits, with regard to structure, exception handling or performance.

The real bottleneck or show stopper is any processing block before the speech signal reaches the speech understanding block, in other words microphones, beam forming, noise reduction, telephone lines (in case of remote access) and the ASR. A poor acoustic front-end or a bad ASR, which results in word accuracy rates below 85% for a keyword spotting recognizer, will lead to a dissatisfying speech dialog system (Krebber 2005).

2.5 Final Check

2.5.1 System Improvement

For most of the speech dialog systems there is enough room for system improvement – every time. In case of INSPIRE majority of the system improvement ideas came after the first set of experiments from users outside the project. In the following is discussed some examples of the INSPIRE system improvements.

Reset patterns
This is a function which allows the system to restart at any given point of the dialog. This feature is usually used when the dialog flow breaks down. It was not implemented in the first version because of structural problems of the general dialog manager.

Generic help vs. context sensitive help
The system help function was redesigned from *universal* or *generic* to *context sensitive*, e. g. whenever the help function is triggered it provides help regarding to the current dialog context (e. g. in case the interaction focus is set to TV, the contents of help will only refer to TV-related issues and it will not present all the possible system functions like in the early system version). The universal help goes back to fundamental decisions made in the system concept. In the end the concept of universal help was rejected after the first set of tests.

Longer prompts vs. shorter prompts
System prompts, which users and developers regarded lengthy, were shortened successfully resulting in a reduction of the average dialog length. There are also slight but noticeable differences regarding the wording of the prompts. Long prompts were created according to the persona of the system. The later change of the prompts required a change in the persona as well. The very polite and formal "servant" was changed to a more friendly and sociable "employee".

Custom actions or short cuts
Custom actions are personalized sequences of an action, for example a function which allows the operation of various devices by a single command, e. g. the possibility of arranging certain default settings which will trigger the actions (e. g. switching off all lamps at once).

Decisions +6 dB
In the first set of tests users got frequently stuck in yes/no decisions e. g. "Would you like to record this movie, yes or no?" They repeated some parts of the sentence like "I would like to record this movie" while the system expected "yes" or "no". By simply boosting the volume by 6 dB of "yes" and "no" in the question, the problem was solved and no participant got stuck at this point during the following experiments. This is a simple but very effective example of subliminal user guidance.

2.5.2 Alternatives

The ultimate question to find something which overrules the usage of a speech dialog system is: "Does the speech dialog system fulfill its aim or can it be replaced?"

For the INSPIRE smart home environments wall mounted switches or remote control handsets may come up as first thoughts.

Wall mounted switches will offer easy access for mobile people, however for disabled people they are out of scope. Remote controls may become easily unmanageable due to the limited surface area. A high amount of buttons results either in small sized buttons or to handle double assigned buttons. Besides, the remote control does not allow remote access via a telephone.

For INSPIRE the universal approach was rated higher by the designers because of the research interest. Nevertheless, other system designers may opt for an additional remote control as fallback solution.

2.6 Basic Advices

In the following some hints and experiences for building a speech dialog system are provided based on the INSPIRE system which a beginner might not find that easily from the literature. Most of the following hints may be taken for granted, however some of them are sometimes overseen during the system design.

- Do not rely on ready applications until you have tested them in your system.
- Room acoustics is still the most underestimated issue for the speech recognizer.
- The telephone transmission lines may vary from extremely good to extremely poor with different coding schemes (usually more than five within one transmission), packet loss, bit errors or noise. Make sure the ASR is robust enough to handle all kinds of signal degradations and test it against some of them for verification.
- Log all the steps the testers are doing. That will help to trace errors.
- Record the interaction with webcams and microphones to separate systems in case of system crashes. Be aware of handling huge amounts of recording data, take care of proper naming, archiving and give an expiration date, as data warehousing may come costly.
- Be aware of the fact that you have to inform your participants that they are recorded and that you get a signed permission from each participant.
- Some laboratories require the permission of the ethic commission, e.g. of the university. Sometimes also journals require the permission of the ethic commission depending on the tests performed.
- Use a professional "listening text" writer, as listening text or "conversational" text for the speech output of a speech dialog system differs quite much from "reading text" or prose.
- Use a professional talker for recording the speech prompts.
- Record the concatenated speech synthesizer prompts as late as possible.
- The longer it takes the user to reach his aim the more problems he has and the less the system acceptance is.
- The more technically advanced these kinds of systems become or the more natural the speech synthesizer sounds, the more the users expect from the system capabilities.
- Expect the unexpected, e.g. losing ready modules because of licensing issues.

2.7   At a Glance

The following bullet points will sum up the fundamental questions before building a smart home speech dialog system.

- Basic requirements and user expectations
  - What is the aim of the system, which shall be the cornerstones?
  - Which is the expected target group using and finally paying for the speech dialog system?
  - Is there a corporate design available?
  - Persona of the system?
  - Which information or action should be provided as the final system action?
  - How good, in terms of quality aspects, does the system have to be?

- Dialog strategies
  - Are all keywords included?
  - How does the dialog system gather all necessary information for a system action?
  - Is the dialog manager capable of handling all expected and unexpected dialog structures?
  - Which communicative situations may appear which need extra caretaking?
  - Would additional input devices and modalities give any benefit, also taking costs for input and output devices into account?

- Technical realization
  - Which components will be part of the system and how do they communicate with each other?
  - How to expand the system?
  - How good, in terms of performance, are single components within the dialog system and which could be the bottleneck or show stopper?

- Final check – system in the loop
  - Is there any chance left for improving the system?
  - Does the system benefit from further modalities?

References

Bernsen, N. O., Dybkjær, H., Dybkjær, L. (1998) Designing Interactive Speech Systems: From First Ideas to User Testing, Springer Verlag, New York

Choukri, K. (1997). System Design. In: Handbook of Standards and Resources for Spoken Language Systems (D. Gibbon, R. Moore and R. Winski, eds.), Mouton de Gruyter, Berlin. pp. 30–78

Docampo Rama, M. (2001) Technology Generations Handling Complex User Interfaces. Doctoral thesis, Technische Universiteit Eindhoven

Fraser, N. (1997) Assessment of Interactive Systems. In: Handbook of Standards and Resources for Spoken Language Systems (D. Gibbon, R. Moore and R. Winski, eds.), Mouton de Gruyter, Berlin. pp. 564–615

Gieselmann, P. (2004) Reference Resolution Mechanisms in Dialogue Management. In Proceedings of the CATALOG Workshop, Barcelona, pp. 28–34

Jekosch, U. (2000) Sprache hören und beurteilen: Ein Ansatz zur Grundlegung der Sprachqualitätsbeurteilung. Habilitation thesis, Ruhr-University, Bochum

Krebber, J. (2005) "Hello – Is Anybody at Home?" – About the Minimum Word Accuracy of a Smart Home Spoken Dialogue System. In Proceedings of EUROSPEECH '05, Lisboa, pp. 2693–2696

ITU-T Recommendation P.851 (2003) Subjective Quality Evaluation of Telephone Services Based on Spoken Dialogue Systems. International Telecommunication Union, Geneva

Möller, S. (2003) Quality of Telephone-based Spoken Dialogue Systems. Springer Verlag, New York

Möller, S., Krebber, J., Smeele, P. (2004) Evaluating System Metaphors via the Speech Output of a Smart Home System. In Proceedings of 8th International Conference on Spoken Language Processing (Interspeech 2004 – ICSLP), Jeju Island. Vol. 3, pp. 1737–1740

Rajman, M., Bui Trung H., Portabella, D. (2004) Automated Generation of Finalized Dialogue Based Interfaces. In Proceedings of Swiss Computer Science Conference SCSC04: Multimodal Technologies, Bern

Trutnev, A., Rajman, M. (2004) Comparative Evaluations in the Domain of Automatic Speech Recognition. In Proceedings of 9th International Conference on Language Resources and Evaluation, Lisboa. Vol. 4, pp. 1521–1524

Trutnev, A., Ronzenknop, A., Rajman, M. (2004) Speech Recognition Simulation and its Application for Wizard-of-Oz Experiments. In Proceedings of 9th International Conference on Language Resources and Evaluation, Lisboa, Vol. 2, pp. 611–614

Chapter 3

Now that we have learned about the need for designing for different user groups – what user groups are there actually? What makes a user group distinct from another one? How can the dialog system know?

Caroline shares some insights on research findings and solution approaches from her work on automatic classification of user groups for speech dialog systems – work that I was given the opportunity to mentor from an industry usability engineer's perspective.

Caroline Clemens, M.A.

After an education in classical singing Caroline Clemens studied Science of Communication and Linguistics with focus on phonetics and speech. As a speech dialog developer she gathered experience in dialog design at Mundwerk AG. Currently she is finishing her doctoral thesis about speech dialog systems supported by Siemens AG, Corporate Technology, User Interface Design. Since 2005 she is a college in the research training group prometei and member of the scientific board of the Centre of Human-Machine-Systems (ZMMS), both at Berlin Institute of Technology.

Dr. Thomas Hempel

Thomas Hempel received his PhD in Communications Research. He worked as a researcher at Bochum University (Germany) conducting studies on industrial sound design, esp. passenger cars. In 2001 he joined Siemens Corporate Technology's User Interface Design Center where he took responsibility for the acoustic peculiarities of user interfaces for both speech and non-speech audio. Now, combining both of his passions acoustics and usability, Thomas is with Siemens Hearing Instruments, Erlangen, Germany.

3 Automatic User Classification for Speech Dialog Systems

Caroline Clemens
Center of Human-Machine-Systems, Berlin Institute of Technology, Germany

Thomas Hempel
Siemens AG, Corporate Technology, User Interface Design, Munich, Germany, now with: Siemens Audiologische Technik GmbH, Erlangen, Deutschland

3.1 Introduction

Although the usability lacks of early speech dialog systems had been discussed in expert circles for quite some years, it took until the 2004 'study on acceptance and usability of speech applications' to make this evident to the German scientific and industrial community. In this study, Peissner et al. (2004) had come up with a systematic and at the same time pragmatic approach to measuring acceptance and usability of existing speech dialog systems. Among others, results show:

- The frequency of use of such systems is very low in German speaking countries.
- Many users are not satisfied by the overall quality of the systems and thus are reluctant to use speech dialog systems at all.
- The success of a dialog system is strongly dependent on the usability of the voice user interface.
- So, from a usability perspective focusing on word recognition rates alone is by far not sufficient to achieve desired user acceptance.

Fortunately, today these issues do not sound too ground-breaking any more and most were known to experts even before. But 2004 these findings had huge impact on the voice business in the German speaking world. It led to re-designs of existing systems and influenced the community in a way to incorporate usability issues at much earlier stages of the system development process. In 2006 a second study states that the usage of speech dialog systems has continued to increase in Germany but also emphasizes the need for further efforts for increasing the usability of such systems (Peissner et al. 2006).

So, whereas in the 1990s technical functionality was the focus, nowadays development processes increasingly concentrate on usability aspects. To guarantee an elaborated dialog design it is necessary to target a prospective system development process from the early development phases on.

Put in technical terms human beings are perfect speech dialog systems of which an important characteristic is that they adapt to their conversation partner. Even in conversations without face to face contact – like telephone calls – people consciously or unconsciously optimize their behavior to the respective dialog partner. In the near future, computers will probably never reach this ability in a comparable performance. This would demand a machine that passes the Turing test. Science fiction computers[1] impress with perfect speech interfaces that allow spoken language conversation like talking to a real person. Some of those have a distinctive personality and are humorous and charming.[2] It might be considered a hypothesis whether they raise popular expectations and make current speech dialog systems seem disappointing but it is a fact that today's systems are not that powerful and cannot copy human communication skills. Concerning this gap between expectation and performance, we detected an interesting effect in our usability tests: If the persona design of a speech dialog system is very realistic (i. e. human like), some users behave in a way that decreases the dialog success. It is known that users who do not respect the limitations of the systems decrease the performance. In our studies, some users overestimated the speech dialog system and expected it to behave like a real person. The recognition problems then increased and the users became disappointed. If this assumption is evidenced one could conclude that it is a disadvantage if a speech dialog system is designed to be too human-like unless it performs correspondingly. The aim should be designing successful dialog systems in terms of user satisfaction, dialog efficiency and matching the user's expectations. Obviously, this cannot be achieved by just humanizing the system.

In our opinion the work of voice interface designers in commercial projects still is underestimated and misunderstood. Reducing their working field to the surface of a dialog system with its audio style and prompt writing is on the same level like saying that graphical user interface (GUI) designers choose nice colors and shapes of buttons. Designing speech dialog

---

[1]   The computer of *Starship Enterprise* or *HAL*, the computer of *2001: A Space Odyssey*, for instance.

[2]   The car *KITT* in *Knight Rider* or *Eddie* on board *Heart of Gold* in Adams' *Hitchhiker's Guide to the Galaxy* series, for instance.

systems concerns the complete design process on all levels. It begins with an analysis of the aims of the dialog, its functions and requirements. When the target group was defined, decisions are made about the target style. Subsequent implementation steps will be based on this, and the architecture, menu and navigation are going to be determined. After that, keywords, audio design, texts, recordings and grammars can be set. For all these steps today designers still rely on their experience and an intuitive estimation of the users. Of course there are books that provide general guidelines (e. g., Morgan and Ballentine 1999; Pitt and Edwards 2002) but it is fruitless to do a literature review in order to find requirements that are characteristic for different types of users. Designers often try to put themselves in the position of a typical user. But who is this typical user? What are her or his attitudes, needs, wishes and preferences? And even more difficult, what about different types of users? Experience shows that there are definitely big differences between users. But how are they characterized? Can this be measured? Novice vs. power user is an established distinction, but again: how are they characterized? Questionnaires (Bearden and Netemeyer 1998; Brunner and Hensel 1994) can be used to observe the user's behavior, but, of course, it is impossible for every user to fill in a questionnaire before using a dialog system.

3.2   Automatic Information Retrieval and Adaptation

In the majority of cases, speech dialog systems are made for a large number of users. In many systems the users are anonymous and do not log in, so there is no advanced information about the user. In these cases a detailed analysis of each individual user by human experts is impossible. Therefore, information about the user and their behavior has to be gathered automatically. Three fields of benefit can be seen for automatic user classification – adaptivity of systems, clues for the designers, and marketing:

Adaptivity of systems
First of all it is needed to clarify the concept of adaptivity, because there are similar concepts like personalization, individualization, adjustment, adaptation and adaptivity. *Personalized* means customized to a specific person. Personalization or *individualization* is only possible if the user is known and identified by the system. This could be done using a password for the login. An example of *adjustment* is a car seat that can be adjusted to suit the user. If a system can adapt itself automatically to modified conditions it is called *adaptive*. *Adaptivity* is the possibility to adjust a system according to the requirements. User oriented systems adapt the system behavior to the user behavior. As there are big differences between

users (heterogeneous audience) it is clear that adaptive dialogs are a great step forward. For this, an automatic detection of user features is necessary. In speech dialog systems, several levels can be adapted to gain a user-friendly dialog, such as content, presentation and interaction style. It has been shown that effects of an automatic adaptation can be higher system performance, more efficiently resolved dialog anomalies, thus higher overall dialog quality (Chu-Carroll and Nickerson 2000; Litman and Pan 1999).

Clues for the designers
Today's research has to focus on the possibilities of building dialogs that fit as well as possible to the actual user. Only a few studies offer helpful design rules for speech dialog systems. Of course there are standards (ISO 9241-110 2006; ISO 9241-11 1998) which contain dialog design guidelines. But such standards are very general and do not help the designer concerning a detailed design decision. If automatic user classification was in place it could automatically derive patterns of use and user interaction and provide it to the designers.

Marketing
Marketing needs to know as much as possible about the target group. Automatic user classification could deliver information in a very economical way. Extracting user features automatically would be of high value for complementing user profiles.

After analyzing existing adaptive dialogs we inspected the technological feasibility of user classification. Although we used telephone-based speech dialog systems, most non-specific characteristics are also relevant for non-telephone-based systems. User interface designers came together in expert workshops to discuss scenarios of automatic adaptation and evaluated estimated benefits. An extract of the results is presented in the following. Sources of information for automatic user classification and adaptation are mainly audio signals, log files and – in case of telephone based dialog systems – the telephone number:

- The audio signal contains the speech signal of the user as well as background noise. Room characteristics and the transmission influence on the acoustic signal. In the speech signal there is information about individual voice, speech and linguistic characteristics like age, gender or accent.
- In a running computer program, log files automatically document when certain program steps take place. They offer the opportunity to register events in a speech dialog system with a precise timestamp. Many events of a dialog are initiated by the user. It is recorded which menu point the

user has chosen at a particular time. There is also documentation of which input is recognized by the speech recognition system or if there is a no-match or no-input. The reaction time of the user can be calculated from the timestamps.

- Where the telephone number is transmitted it represents an information source. The number consists of a prefix and the extension. If the call is coming from a foreign country the country prefix is transmitted, too. The prefix is either an area code or mobile network number. So you know whether the user has called from a landline or a mobile phone. The area code tells you which region the user is calling from or the mobile network number tells you which mobile network is being used. It is technically possible to detect how often and at what times a certain telephone number is calling the speech dialog system.

The following examples show how the dialog system can adapt to automatically retrieved information. All adaptations suggested here are technically feasible and some of them have already been realized, such as a classification of gender and age (Stegmann et al. 2006, Metze et al. 2007) and adaptations to the user's observed level of expertise (Jokinen 2006).

Age:
Existing automatic age classifiers can differentiate between broad age groups. In languages that use different forms of address like German ("Du"/"Sie") or French ("Tu"/"Vous") the form used by the system could be fitted to the age group of the user. Speaking and language style as well as target group focused advertising could be chosen.

Gender:
There are classifiers that can detect the gender of a speaker with a high success rate. According to the detected gender the system could be adapted adequately. For instance, if it is known that men and women prefer different speakers then a suitable speaker could be chosen.

Car:
An acoustic analysis can detect if the speaker is in a car. While driving, the user might get distracted by the driving task and not react immediately. Then the timer could allow longer reaction times for a no-input time out.

Emotions:
Emotions such as anger or frustration can be detected automatically (Ang et al. 2002). The system could react by transferring to an operator and thus avoid a hang up.

Native language:
There are techniques to find it out which language a user is speaking in and the dialog system could switch to this.

Number of barge-ins:
Many barge-ins that do not produce no-matches can indicate an experienced power user. The system could react with a faster navigation, shorter prompts and fewer explanations.

Timestamp of barge-in:
Early barge-ins show that the power user knows the next input before the prompt is played completely. Shortcuts could be offered to allow a shorter dialog for the next call.

Number of no-matches or no-inputs:
If many no-matches are produced the user may need more explanations. It might be helpful to reduce the input possibilities to keywords in order to improve the recognition rate. If recognition is still problematic, the user could be transferred to an operator.

Path and pattern of navigation:
If a user wanders around in the navigation, the system could help with declarations and a more directed dialog.

Number of help requests:
A large number of help requests shows that the user needs more declarations and a more directed dialog style.

Telephone extension number:
Storing telephone numbers in a database allows checking whether a number has called before, how often, and at what time of day. Extensions of some companies or public authorities all begin with the same numbers. When it comes to saving data, privacy has to be respected. Of course the same telephone number does not guarantee the same user which makes it necessary to handle sensitive data responsibly.

Telephone prefix:
If the prefix is an area code, local services could be offered for the location the user is calling from. The prompt texts could be chosen in accordance with local conventions.

To supplement the results of the expert examination two focus groups were organized with five members each. After a short introduction to telephone based speech dialog systems, there was time for a moderated discussion about possible adaptations of such a system. The results of the two focus groups confirmed those of the expert examination. The focus groups delivered some additional ideas but showed up the same critical points and no really new aspects. After both the expert and the focus group studies were examined we decided to focus on log files in further studies, because as a source of information they offer possibilities that are easy to realize and powerful at the same time. Other sources of information are either too complex and therefore too expensive to develop or their automatic classifiers are still not mature enough for state of the art performance.

3.3   User Features and Log Files

The following list shows a clustered collection that describes the wide range of personal, cultural and situational factors that influence the user's behavior:

- situation, location, attendant persons, time pressure, other activities and tasks beside using the system itself (e. g. driving as major task);
- demographic factors like age, gender, native language;
- experience, knowledge, skill, expertise, intelligence, cognitive workload, memory decline, attention, capability, expectation, preferences, habits, interests;
- personality, traits, type, character, attitude;
- mood, health, emotions, stress, condition, state;
- the person's task and aim of use;
- the speech dialog system itself.

Determining all these factors is impossible. It is important to know how the user behaves and how the dialog can adapt to that behavior in the best way. The features can be grouped in four dimensions: *User features* concern the user and the user's interaction features; *Task features* picture the task of the concrete use case; *Technical features* are set by devices and infrastructure used for solving the task; *Surrounding features* like situation, setting, local and temporal context.

Given our focus, user features have to be extracted from parameters that are retrieved from log files. Depending on the architecture of the system there is more than one module that creates a log file like the speech recognition system or the dialog management server. If several log files have to be combined it is important to assure synchronism. Typically, log file entries will have a form like [timestamp|event|detail1|detail2|...]. Events that are logged are basically: Start of the dialog, switches to sub-dialogs or menu points, events of the recognition system including recognition rate and recognized words, and end of dialog. From the logged dialog events, parameters have to be extracted that can describe the user's interaction:

- Dialog duration;
- User turn duration;
- User response delay;
- Number of user turns;
- Words per user turn;
- Number of help requests;
- Number of time-out prompts;
- Number of ASR rejections;
- Number of error messages;
- Number of barge-ins;
- Number of cancel attempts.

Interaction features extracted from log files have already been used to predict the quality of a dialog system (Möller 2004). To make the interaction features more comprehensible they can be aggregated:

Speed:

- Reaction time of user in prompts;
- Mean duration of inputs;
- Call duration;
- Difference between the length of the user's path in the dialog and a defined optimal path for solving the task.

Competence:

- Use of barge-ins;
- Number of no-inputs;
- Number of no-matches;
- Number of help requests;
- Goal-oriented navigation.

Cooperativeness:

- Use of proposed keywords;
- Number of words per utterance.

Learning effect:

- Decreasing number of no-inputs;
- Decreasing number of matches.

Politeness, conformance with conversation rules of human-human dialogs:

- Number of barge-ins;
- End of call: leave-taking/good-bye.

3.4   Testing and Findings

A preceding usability test was conducted in cooperation with T-Systems Enterprise Service GmbH, Berlin, Germany. We examined differences between users of different gender and age, resulting in the user groups: adult male, adult female, senior male, senior female, and children. After a questionnaire about demographic data, the test person made calls with a Wizard of Oz speech dialog system. This was a fictitious timetable information system for public transport that included an automatic age and gender classifier. After the test calls, the user's opinions and experience were observed in guided interviews.

The test showed a higher drop-out rate for senior test persons. The problematic or unsuccessful calls showed noticeable characteristics like a high number of no-matches, long inputs with many words in long phrases and few keywords only. No noticeable learning effect could be observed and the interaction of the test person conformed strictly to human-human interaction conventions. Additionally, those test persons appeared nervous, aroused or frustrated, some reported they felt misunderstood by the system.

All user groups in the test preferred a more directive dialog style in contrast to mixed-initiative dialogs. The more calls the test persons made the more target oriented they performed. This learning effect is reflected in lower error rates.

The automatic classification of gender and age to the user groups was correct in 56% of the cases. All wrong classifications regarded the age of the test person whereas all gender judgments were right.

In our main study, data was collected from test persons in three age groups (children 9–14 years, adults 20–45 years, seniors 63–72 years). Additionally, personality related data were gathered. We retrieved log files of around 12 calls per person with different tasks and scenarios as well as additional material such as video and audio recordings and notes of the test leader.

The telephone-based speech dialog system we used included automatic speech recognition for spoken language input and played recorded audio files as acoustic output. Common global navigation keywords were active and barge-ins were accepted. The current state[3] of analysis points out the following issues:

- The collected data indicates that there is a relation between users' execution speed and their overall technical experience.
- Increasing experience with the dialog system under test leads to changes in some users' behavior (learning effect).[3] The common terms novice and power user are in fact useful.
- There is a relationship between age and learning effect.
- There is a bigger variance in the results of the group of seniors than in the group of younger adults.
- There are more elderly with low tempo and little technical experience.
- There is a greater variance in performance of users with little or no experience compared to the variance among experienced users.

3.5   Summary and Outlook

The usability of a speech dialog system depends crucially on the quality of the dialog design. For obtaining high user acceptance for the system, the user centered dialog design should already be focused on during the early phases of the system development process.

From the perspective of our studies the usage of adaptive speech dialogs is a big technological and creative step since this enables both new user-centered speech dialog designs and new marketing opportunities.

Furthermore, automatic classification and analysis data is about to lead to yet undiscovered relations within measurable parameter sets in users' interaction behavior. Therefore, it is important to collect the named interaction features during the dialog – and based on such results research is about to develop more advanced classifiers.

Also, it is important to determine details like the time needed by a classifier to reach a reliable classification result during a user's call. But even if it takes too long technically to be able to use the result in the current user dialog, the result could be stored and be available for the next call from the same telephone number.

There might be more detectable patterns in the navigation behavior of users. Recurring navigation patterns like loops, jumps or iterations in the dialog flow can correlate with known user features. But even in some years

---

[3]   Parts of the analysis are still pending. For further findings and a detailed description of procedure and results see first author's doctoral thesis (to be published 2008).

when automatic classification is expected to work even more reliable, voice dialog designers will face their true challenge of how to adapt the dialog to the satisfaction of the respective user group. Results of user tests of the different user groups have to be carefully analyzed. Only then the results could help to formulate adequate style guidelines for the design of better – user focused – dialog systems.

Acknowledgements

The presented studies were conducted within a research project of Siemens AG, Corporate Technology, User Interface Design in cooperation with the Center of Human-Machine-Systems at Berlin Institute of Technology. They were part of a project on Adaptive Speech Dialogs funded by Siemens COM's Chief Technology Office. The project was part of a cooperation of Siemens AG with Germany's Deutsche Telekom AG. The authors would also like to thank Siemens CT project manager Dr. Ing. Bernt Andrassy (Professional Speech Center, www.siemens.com/speech) and Frank Oberle and his team at T-Systems Enterprise Service GmbH, Berlin, for the fruitful co-operation.

References

Ang J, Dhillon R, Krupski A, Shriber, E (2002) Prosody-Based automatic detection of annoyance and frustration in human-computer dialog. In: Proceedings of the 7th ICSLP (International Conference on Spoken Language Processing), Denver

Bearden WO, Netemeyer RG (1998) Handbook of Marketing Scales: Multi – Item Measures for Marketing and Consumer Behavior Research. 2nd ed. Newbury Park, Sage Publ., Calif

Borkenau P, Ostendorf F (1993) NEO-Fünf-Faktoren-Inventar (NEO-FFI) nach Costa und McCrae. Hogrefe, Göttingen, pp 5–10, 27–28

Brunner GC, Hensel PJ (1994) Marketing scales handbook: A compilation of multi-item measures. American Marketing Association, Chicago

Chu-Carroll J, Nickerson JS (2000) Evaluating automatic dialogue strategy adaptation for a spoken dialogue system. In: Proc. NAACL 1, pp 202–209

ISO9241-110 (2006), Ergonomics of human-system interaction – Part 110: Dialogue principles

ISO9241-11 (1998), Ergonomic requirements for office work with visual display terminals (VDTs) – Part 11: Guidance on usability

Fahrenberg J, Hampel R, Selg H (2001) Freiburger Persönlichkeitsinventar (FPI). Revidierte Fassung (FPI-R) und teilweise geänderte Fassung (FPI-A1). Hogrefe, Göttingen

Jokinen K, (2006) Adaptation and user expertise modelling in AthosMail. In: Universal Access in the Information Society 4 pp 374–392

Litman DJ, Pan S (1999) Empirically Evaluating an Adaptable Spoken Dialogue System. In: Proceedings of the 7th International Conference on User Modeling

Metze F, Ajmera J, Englert R, Bub U, Burkhardt F, Stegmann J, Müller C, Huber R, Andrassy B, Bauer JG, Littel B, Comparison of Four Approaches to Age and Gender Recognition for Telephone Applications, ICASSP 2007. IEEE International Conference on Acoustics, Speech and Signal Processing, 2007, pp IV-1089–IV-1092

Morgan DP, Balentine B (1999) How to Build a Speech Recognition Application: A Style Guide for Telephony Dialogues, Enterprise Integration Group, San Ramon

Möller S (2004) Quality of Telephone-Based Spoken Dialogue Systems. Springer-Verlag, Berlin

Peissner M, Biesterfeldt J, Heidmann F (2004) Akzeptanz und Usability von Sprachapplikationen in Deutschland. Technische Studie, Fraunhofer-Institut für Arbeitswirtschaft und Organisation (IAO), Stuttgart

Peissner M, Sell D, Steimel B (2006) Akzeptanz von Sprachapplikationen in Deutschland 2006, Fraunhofer-Institut für Arbeitswissenschaft und Organisation (IAO), Stuttgart und Initiative Voice Business, Bad Homburg

Pitt I, Edwards A (2002) Design of Speech Based-Devices. Springer-Verlag, New York

Stegmann J, Burkhardt F, Oberle F, Eckert M, Englert R, Müller C (2006) Einsatz der Sprecherklassifizierung in Sprachdialogsystemen. In: Tagungsband der 7. ITG-Fachtagung Sprachkommunikation, Kiel

Chapter 4

Sebastian Möller and his renowned project partners take things a step further: How can automatically obtained user information be exploited in a tool that helps optimize future speech dialog systems by considering as much information as possible from a user perspective?

Users' assessments from real usability tests were taken and compared with data collected in the system's log files.

Prof. Dr.-Ing. Sebastian Möller

Sebastian Möller is Professor for Usability at the Quality and Usability Lab, Deutsche Telekom Laboratories, Berlin Institute of Technology, Germany. He received his PhD (Dr.-Ing.) from Ruhr-Universität Bochum in 1999 for his work on the assessment and prediction of speech quality in telecommunications, and got the qualification to be a professor (venia legendi) at that university in 2004, with a book on the quality of telephone-based spoken dialogue systems. His main research interests are on speech processing, speech technology, communication acoustics, quality assessment, and on the usability of telecommunication services. Since 1997, he has taken part in the standardization activities of the International Telecommunication Union (ITU-T).

Klaus-Peter Engelbrecht

Klaus-Peter Engelbrecht is working as a research assistant at the Quality and Usability Lab of Deutsche Telekom Laboratories (T-Labs), TU-Berlin. He studied Communication Research and Musicology and received his Magister degree in 2006 from Berlin Institute of Technology. At T-Labs, he is working towards his PhD thesis in the domain of automated usability evaluation for spoken dialog systems.

Dr. Michael Pucher

Michael Pucher studied philosophy at the University of Vienna and computational logic at Vienna University of Technology. Since 2001 he is working at the Telecommunications Research Center Vienna (ftw.) where he is currently senior researcher and project manager. In 2007 he received a doctoral degree in Electrical Engineering from Graz University of Technology. He was a visiting researcher at the International Computer Science Institute in Berkeley (ICSI) and at Deutsche Telekom Laboratories in Berlin (T-Labs). His research interests are speech synthesis, language modeling for speech recognition, and multimodal and spoken dialog systems. Currently he is working on Viennese sociolect and dialect speech synthesis for regionalized voice services.

Dr. Peter Fröhlich

Peter Fröhlich studied Psychology at University of Salzburg and music education at Mozarteum Salzburg. In 2007, he received a doctoral degree in Applied Psychology from University of Vienna. From 2000 to 2003, Peter worked as usability specialist at CURE – Center for Usability Engineering and Research in Vienna. Since December 2003, he joined Telecommunications Research Center Vienna (ftw.) as a Senior Researcher responsible for the Human-TelecomSystems Interaction Lab. Peter is also lecturer for Usability at FHWien University of Applied Sciences. His research interests include auditory and multimodal HCI, research methodology, mobile spatial interaction, eLearning, and interactive television.

M. Sc. Lu Huo

Lu Huo was born in Wuhan, China in 1979. She received the Master of Science (M.Sc.) degree in digital communication from the Christian-Albrechts-University of Kiel, Germany, in 2006. Since April 2006 she is with the Institute for Circuit and System Theory, University of Kiel, as a Research Assistant. Her research interest is the instrumental assessment of wideband-speech quality and TTS (Text to speech) quality.

Prof. Dr.-Ing. Ulrich Heute

Ulrich Heute was born in 1944. He studied Electrical Engineering in Stuttgart and received his PhD in 1975 in Erlangen. Working at Erlangen University for some years as a senior researcher, he received his Habilitation in 1983. In 1987, he became a Professor for Digital Signal Processing at Bochum University (Germany). Now he is Head of the Institute for Circuit and System Theory at Kiel University (Germany). His research interest comprises digital filters, filter banks and spectral analysis for as many applications as audio, sonar, radar, and ECG/EEG/EMG signals. Currently he is focusing on speech-signal processing (medium-to-low rate coding, noise reduction, speech-signal models, speech-quality measures, wide-band speech, voice conversion).

Dipl.-Ing. Frank Oberle

Frank Oberle is consultant in the field of Innovative Voice Solutions at Deutsche Telekom's system integration division T-Systems. He is leading the working field "Innovative Application Design" and is responsible for the conception of tools for the speech application development at T-Systems. Focal point of his work is the co-operation in innovation projects of the Deutsche Telekom, focused on the integration of new technologies into speech dialog systems and the conception of innovative, multimodal solutions. He studied electrical engineering at the Technical University of Berlin and has a longstanding experience in language technology and speech application development.

4 A new Testbed for Semi-Automatic Usability Evaluation and Optimization of Spoken Dialogue Systems

Sebastian Möller[1], Klaus-Peter Engelbrecht[1], Michael Pucher[2], Peter Fröhlich[2], Lu Huo[3], Ulrich Heute[3], Frank Oberle[4]

[1] Quality and Usability Lab, Deutsche Telekom Laboratories, Berlin Institute of Technology, Germany;
[2] Forschungszentrum Telekommunikation Wien (ftw.), Austria;
[3] LNS, Christian-Albrechts-Universität Kiel, Germany;
[4] T-Systems Enterprise Services GmbH, Berlin, Germany

4.1 Introduction

As it has been pointed out in the first part of this book, the development of a usable, high-quality spoken dialogue service requires the development cycle to be *user-centered*. Demands and expectations of potential users have to be anticipated early in the design process in order to build systems which fulfill or even exceed these demands and expectations. Determining the degree of demand fulfillment, however, is not an easy task: Members of the target user group have to be invited to the usability laboratory, confronted with prototypical tasks in an interaction experiment, observed with respect to their interaction behavior, and finally asked for their opinion about the interaction experience. Unfortunately, measurement of quality and usability is not yet possible without the direct involvement of human test participants, acting as perceiving and judging "measurement organs" [1].

On the other hand, commercial spoken dialogue telephone services are usually implemented on dialogue platforms that are able to log a multitude of information upon request. Such information includes the audio signals recorded from the user, the signals played to the user, the recognition and understanding results, or information on the state of the dialogue machine. Because of its large amount and diversity, this information is very difficult to interpret. However, since the cues of dialogue flows that are experienced by the user, and thus the cues of the perceived quality and usability, might be hidden in this log information, analysis of the log data could be very useful for quality and usability analysis, optimization, and monitoring.

In order to take full profit of log data, their relationship to perceived quality and usability has to be estimated. This relationship is not straightforward. For example, a system developer may initially assume that a better coverage of the users' utterances by the recognizer's vocabulary will decrease the frequency of system non-understanding instances and increase dialogue efficiency and user satisfaction. However, a larger vocabulary will also reduce the overall recognition performance, and may finally be detrimental to dialogue efficiency and user satisfaction. Apparently, there is no simple one-to-one relationship between vocabulary coverage (or system characteristics in general) and user satisfaction (or user perceptions in general). In order to find an optimum setting for a specific system and design problem, we need to know

- the impact of each change of the system on the interaction behavior, and
- the relationship between interaction behavior and perceived quality and usability.

Provided that both relationships are known, it is possible to estimate the impact of each system characteristic – and each change in one of several characteristics of a complex system – on perceived quality and usability.

Making use of log-file information and combining them with prior knowledge of the system to predict quality and usability will largely accelerate the design and implementation cycle, and (surprisingly) will turn it more *user-centered*: If information on the behaviors of system and user is available without directly asking the user, and without the need for an expert transcription and annotation, and if quality and usability can be estimated on the basis of this information, the user's demands can be taken into account in *all* stages of system development, *without additional costs*. For example, the system developer can estimate the consequences of each modification of her design with respect to the ultimate target – getting highly satisfied users. The log data at the basis of the estimation can be obtained e. g. by offering the prototypical service to a limited number of "friendly users"; there is no further need for a formal usability test at this stage of the development cycle.

In this chapter, we make use of information which can easily be logged by state-of-the-art dialogue platforms for analyzing, optimizing and monitoring the quality of spoken dialogue interactions. Our focus is on parameters which can be extracted automatically, without requiring a tedious and time-consuming transcription or labeling process. We further limit our analysis to telephone-based services, which form a large class of current commercially-deployed services.

We extract information both on the signal and on the symbolic level. The algorithms used for this purpose are described in Sect. 4.2. With the help of these algorithms, we analyze data collected within a usability test of a prototype implementation for a pre-qualifying application of Deutsche Telekom AG, as it is described in Sect. 4.3. The test was performed in cooperation with Siemens AG, Corporate Technology, Competence Center "User Interface Design". The extracted parameters are analyzed in Sect. 4.4, first by addressing their relationship to the recognition performance of the system, and then to the subjective judgments obtained from the test users. We finally describe how the algorithms are integrated into a graphical testbed for an easy and efficient log-file analysis in Sect. 4.5. We conclude and give an outlook on possible use cases of the testbed in Sect. 4.6.

The chapter extends previous work described in [2] and [3] and includes previously unpublished results obtained within the TIDE project funded by Deutsche Telekom AG [4].

4.2 Basic Information Related to Quality and Usability

Different types of information may be useful for addressing the quality and usability of spoken dialogue services. This information is either part of the *system characteristics* (which usually remain stable during a particular interaction, except when dealing with adaptive systems), or it evolves when users interact with the system (*interaction behavior*). The latter may be obtained on a *signal* or on a *symbolic* level.

The primary information exchanged between user and system is reflected in the audio signals from the user and the system. These audio signals are usually available on the system side, both on the input (originating from the user and directed to the system) and the output side (originating from the system and directed to the user). Because the system is not always "listening", i.e. the speech recognition channel is not always open, the audio signal recorded from the user may be influenced by the voice activity detection (VAD) algorithm implemented on the platform. It may also be enhanced by noise and echo suppression algorithms. The system's speech signal is commonly available as a clean pre-recorded audio file, or as a clean file generated by a text-to-speech (TTS) module.

The interaction between user and system can additionally be quantified on a symbolic level, in terms of interaction parameters. These parameters describe the behavior of the user and the one of the system, and reflect the performance of the system in the interaction. Examples include the number and length of user and system utterances, timing information, the frequencies of help requests, cancel attempts, or barge-ins, recognition and understanding accuracies, or task success parameters. In ITU-T Suppl. 24 to P-Series Recommendations [5], a large number of such interaction parameters are described. However, the determination of most of these parameters requires a manual transcription and annotation by a human expert.

Some information is also hard-coded in the system, e. g. in terms of the interaction logic, the vocabulary, or the grammar which is available at each state of the dialogue. These system characteristics have a significant impact on the dialogue flow and on the quality of the interaction, as it is perceived by the human partner. Quality, however, can only be determined by collecting subjective judgments in controlled experiments. ITU-T Rec. P.851 [6] describes methods for carrying out such subjective interaction experiments. They are usually very expensive, and as a consequence, system designers try to avoid them as far as possible.

In the following sub-sections, we describe which information has been extracted for the purpose of our experiments described in Sect. 4.3, both on the signal and on the symbolic level. This information will form the basis for experimental analyses and quality predictions in Sects. 4.4 and 4.5.

4.2.1 Information on the Signal Level

The audio signals that have been recorded at the dialogue platform in our experiments include

- the user's speech signal, transmitted through the telephone channel and cut out by the VAD, and
- the system's speech signal available as a clean audio file.

No audio signals are available on the user's side. Thus, we do not have any information on how the system's speech signal is degraded by the transmission channel before it is perceived by the user, nor on the clean speech signal uttered by the user.

Based on the observations of the database described in Sect. 4.3, the user's speech signal is analyzed with respect to the characteristics of the user as well as of the transmission channel. For this purpose, the following parameters are extracted:

1. *Active Speech Level (ASL)*: This level is calculated from the segments extracted by the VAD, following the algorithm described in ITU-T Rec. P.56 [7]. It consists of a histogram analysis with multiple variable thresholds and results in a level relative to the overload point of the digital system.
2. *Noise Level*: Noise mainly stems from background noise present at the user's side and picked up by the telephone handset, as well as from circuit noise induced by the subscriber line. Two algorithms have been compared to determine the noise level: (a) Speech pauses have been extracted with the help of the GSM VAD [8], and then a smoothed noise

power spectrum is determined from the windowed non-speech segments. (b) Alternatively, minimum statistics [9] has been used to extract
the noise power during speech segments. Method (b) tended to overes-
timate the noise level on a set of controlled test data, and therefore we
decided to use method (a) for noise level determination.

3. *Signal-to-Noise Ratio (SNR)*: With a similar processing as for the noise
power, a smoothed power spectral density of the speech signal is determined during speech activity. The SNR is calculated as the ratio between both power densities, calculated per utterance.

4. *Mean Cepstral Deviation (MCD)*: It is known that multiplicative noise
is introduced in telephone channels by logarithmic Pulse-Code Modulation (PCM) or Adaptive Differential Pulse-Code Modulation (ADPCM)
coding. In order to determine the level of degradation introduced this
way, we assume that the recorded speech signal $y(k)$ is determined by
the clean speech signal $s(k)$ and a white Gaussian noise component $n(k)$
with a certain ratio Q:

$$y(k) = s(k) + s(k) \cdot 10^{-Q/20} \cdot n(k) \tag{1}$$

Falk et al. [10] proposed to measure the noise in the degraded speech
signal via the flatness of the output speech signal $y(k)$. The underlying
idea is that – because the multiplicative noise of Eq. (1) introduces a fair-
ly flat noise in the spectral domain – the lower the Q value, the less the
spectrum of $s(k)$ can be preserved in the noise output $y(k)$, and the flatter
the spectrum of the noisy output is. We use the MCD as a measure of the
amount of multiplicative noise present in the degraded speech signal,
because analyses have shown that the correlation between Q and MCD
is about −0.93 [10]. To calculate MCD, we determine speech segments
with the help of the GSM VAD, calculate cepstral coefficients for the
speech frames, and average the standard deviations of the cepstral coefficients.

5. *Single-ended Speech Quality Estimate*: Multiplicative noise is not the
only degradation introduced by modern telephone channels. In particular, non-waveform speech codecs generate distortions which have different perceptual and signal correlates, and which have shown to de-
grade recognition performance [11]. In order to cover these channel
degradations, we used the single-ended model described in ITU-T Rec.
P.563 [12] to obtain an indication of the overall speech quality degradation introduced by the channel. This model generates a clean speech reference from the degraded speech signal by means of an LPC analysis
and re-synthesis. Both the generated clean and the recorded degraded
speech signals are transformed to a perceptually-motivated representation. An estimate of the overall quality, MOS, is then determined from

a comparison of both representations. The approach can also be applied to TTS signals generated by the system, as it has been shown in [13].

6. *Active Speech Duration*: We use the GSM VAD to cut off pauses at the beginning and at the end of the speech signals which remain after the VAD of the dialogue platform, probably to comfort the speech recognizer.

7. *Fundamental Frequency (F0)*: Hirschberg et al. [14] have shown that mean and maximum *F0* can be useful predictors for the recognition error. We adopted the autocorrelation analysis from Rabiner [15] and some simple smoothing algorithm in order to obtain reliable *F0* estimates. From the *F0* contours obtained for each user utterance, we calculated the mean, the standard deviation, and the 95% percentile.

Parameters according to 1–5 mainly reflect the characteristics of the telephone transmission channel, whereas 1, 6 and 7 address the characteristics of the user. Thus, by determining these parameters, it may be decided whether recognition failures are due to the transmission channel or to user particularities.

4.2.2   Information on the Symbolic Level

Apart from the audio signals, the dialogue platform logs information related to the dialogue state the system is in, as well as to the speech recognizer. For the dialogue state, the system logs the ID of the respective state, a time stamp when the system enters the state, the prompt which is played in this system state, as well as the ID of the vocabulary and the grammar used by the speech recognizer in the state.

The vocabulary of the system has been analyzed with respect to its confusability, as this is expected to be predictive of the recognition performance of the system in that specific state. We calculate four types of confusion measures: (1) The Minimum Edit Distance (MED) with an equal weight for substitutions, insertions and deletions; (2) an articulatory phonetic distance, where the substitution costs of the MED have been weighted according to an articulation-based phonetic distance; (3) a perceptual phonetic distance, where the MED has been weighted according to perceptual similarities; (4) an HMM distance based on the Kullback-Leibler divergence between two Gaussian mixture models. Details on the measures are given in [3]. They show that methods 2–4 all provide reasonable correlations with word confusion. Because the acoustic models of the speech recognizer are not accessible on the dialogue platform, only the articulatory (2) and the perceptual phonetic distances (3) are discussed here.

The speech recognizer of the dialogue platform provides a status flag indicating "recognition" (correct or incorrect), "rejection" (due to the rejection threshold of the system), "hang up" (the user hangs up), "speech too early" (the user speaks too early), "no speech timeout" (the user did not speak in the time interval where the recognizer was open), or "leftovers" (the user leaves the dialogue without speaking). The labels "recognition" and "rejection" refer to the behavior of the recognizer and can be evaluated with respect to the recognition performance.

4.3   Data Acquisition

The described algorithms have been applied to data collected with a telephone-based system for telephone tariff information. This system is implemented on a Nuance dialogue platform. It provides information on fixed and mobile telephone and internet tariffs, and allows internet problems to be reported. In addition, callers are classified according to their age and gender, and the system persona is adapted accordingly. Personas differ with respect to their voice, prompt style and form of address. The system generates log-files and records the user's speech signal in each dialogue state.

25 native German test subjects interacted with a prototype implementation of this system. The subjects were recruited according to 5 groups: AF (adult female), AM (adult male), SF (senior female), SM (senior male), and C (child), where seniors were 60–80 years old, adults were between 25–45 years of age and children were 9–13. All subjects had to carry out 12–13 interactions with the system, targeting on fixed telephone tariff inquiry, mobile telephone tariff inquiry, internet tariff inquiry, and internet problem report. In some of the dialogs, a wrong system persona was chosen deliberately to examine the consequences of user classification errors. As some senior and child participants were not able to complete all tasks, the experiment resulted in a set of 280 dialogues and 1672 user audio files in the database.

After each interaction, the users rated a questionnaire with several items related to their current experience. Here we consider user judgments on the four general items

- overall impression (1...very good – 6...unsatisfactory),
- system wording comprehensible (1...yes – 5...no),
- system understood what user wanted (1...yes – 5...no),
- dialogue should be changed (1...yes – 2...no),

which have been collected after each dialogue.

4.4   Data Analysis

All user audio files have been transcribed by a human expert. The transcriptions and the recognizer's hypotheses have been compared with the help of NIST's "sclite" software [16], determining the number of correct words, substitutions, deletions and insertions for each user utterance classified with the "recognition" or "rejection" label.

Table 4.1. Summary of user speech characteristics

Parameter	Mean	STD
ASL (dB)	−24.5	7.8
Noise level (dB)	−57.4	7.9
SNR (dB)	32.1	13.3
MCD	0.103	0.008
Single-ended estimate	2.46*	0.85*
Active speech duration (s)	1.53	1.87
F0 mean (Hz)	165.0	45.2

*) Due to the short utterance length, P.563 estimations had to be derived from artificially concatenated segments per dialogue, which might have caused low MOS values.

A general analysis of all user utterances is given in Table 4.1. It shows that the utterances have a high signal-to-noise ratio and are of relatively high speech quality. We think that this is due to the test set-up where subjects interacted with the system from two test cabinets equipped with a good wireline telephone. In addition, the utterances are relatively short, indicating that the subjects preferred to use a simple command language towards the system.

4.4.1   Analysis with Respect to User Groups

An analysis by user group was carried out for parameters related to the signal quality (ASL, noise level, MCD, recognizer status labels). Data was aggregated on a dialogue level (mean values for utterance-wise variables) and ANOVAs were calculated taking into account that each test participant contributed 3–13 cases to the data set. As was expected, the signal quality parameters do not differ between participant groups, while the label "recognition" shows a significant effect (ANOVA parameters $F = 3.9$; $p = 0.02$), which however does not hold for the label "rejection".

Groups differed significantly with respect to their commanding style as parameterized in active speech duration, timeout and barge-in frequency ($F{>}4.3$; $p{<}0.01$). SF (senior female) show highest means for duration and barge-ins and directly follow C for timeouts, while M, F and SM have comparatively homogeneous means for duration and timeout.

Interestingly, the judgment on overall impression differs significantly depending on the groups ($F = 8.4$; $p = 0.00$), SM rating worst, followed by SF, M, F and C. The judgment on system understanding shows the same tendency, however, SF and F each gain one rank, which is in accordance with the observed recognition performance being lower for F than for M and for SF than for SM.

4.4.2 Signal Level Parameters and Recognition Performance

In order to represent recognition performance, the recognition results are classified into two classes "correctly recognized" and "error", where the first term refers to the complete match between the transcript of the user signal and the recognition result and the second refers to any mismatch or rejected situation. Table 4.2 summarizes the ASR performance for different user groups.

Table 4.2. Summary of ASR performance for different user groups

User group	No. of recognized utterances	No. of rejected utterances	No. of correctly recognized utt.	% of correctly recognized utt.
Adult male	295	41	273	81.25
Adult female	248	52	225	75.00
Senior male	193	82	166	60.36
Child	131	63	119	61.34
Senior female	97	92	85	44.97
All	964	330	868	67.08
All but senior female	867	238	783	70.86

The user group SF experienced most of the recognition failures because they tended to use complex and long sentences; because of this, we decided to exclude this user group from further analysis. After exclusion, 29.14% of the 1105 remaining recognition results are classified as "error". So without any parameters in the prediction model, we can predict the recognition error with an error rate of 29.14%. Any model that reduces this error rate can help to predict the recognition performance.

Inspired by the work of Hirschberg [14], we adopt the rule-learning program "RIPPER" from Cohen [17]. "RIPPER" is used here to generate plausible rules based on the extracted parameters from the speech signal that can be used to predict ASR performance. Figure 4.1 is an example of the resulting rules from "RIPPER" that can decrease the prediction error from the baseline 29.14 % to 21.63%.

```
if ACTIVE_DURATION>=1.39 && MCD>=0.11,
    then Error.
if ACTIVE_DURATION>=1.82 && SNR>=-51.30 && MCD>=0.098 &&
    MCD<=0.11 F0_MEAN>=129.92,
    then Error.
if ACTIVE_DURATION>=0.92 && F0_MEAN<=105.14,
    then Error.
else Correct.
```

Fig. 4.1. Rule set for predicting recognition errors. The prediction error is 21.63% +/- 1.24%.

The parameters used in the prediction model are selected using the "stepwise" method: At first, we just select one of the nine candidate parameters into the prediction. Then, more parameters are added to the model one by one if they decrease the prediction error. Table 4.3 contains the selected best results for the prediction model with 1, 2, 3, 4 and 5 parameters, respectively.

From these results, it seems that the active speech duration and multiplicative noise are the most significant parameters in the prediction model using our database.

Table 4.3. Prediction error for predicting ASR performance

Model parameters	Error %	Model parameters	Error %
Active duration, MCD, Noise level, ASL, SNR, Mean F0	20.69	Active duration and MCD or Active duration and ASL	22.71
Active duration, MCD, SNR, Mean F0	21.18	Active Duration	25.52
Active duration, MCD, ASL	22.26		

4.4.3 Vocabulary Characteristics and Recognition Performance

The three distance measures (1) Minimum Edit Distance (MED), (2) MED with articulatory phonetic features, and (3) MED with perceptual phonetic features were optimized on held-out word confusion data. We did not use the HMM distance here, because we had no access to the acoustic models that generated the test data. The optimized measures were used to evaluate the correlation between word-confusion and word-distance, and between interpretation-confusion and interpretation-distance.

The test data for this evaluation consisted of 1091 utterances from our spoken dialogue system, with a Word-Error-Rate (WER) of 16.4%. The understanding error rate was however only 10.9%. This error rate was determined by comparing the natural language (NL) interpretation of the reference utterance with the one of the recognized utterance. If a gold standard for interpretations is applied, then the understanding error rate is only 6.5%. A gold standard is a mapping between word strings and interpretations, e. g. between syntax and semantics, which determines the correct semantic interpretation of a string of words in a certain context. Up to this day, such a standard can only be established through manual labeling, which we did not perform for our test set.

Table 4.4. Example of word errors and understanding errors

	Words	NL interpretation
Recognized	NACHFRAGE zu aufträgen	target-aufträge
Reference	NACHFRAGEN zu aufträgen	[no interpretation]
Recognized	ABRECHNUNG	target-rechnung
Reference	ABBRECHEN	abbrechen

To evaluate the performance of the measures for the test set, we estimated the correlation of word confusion and phonetic distance of words. Additionally, the correlation of natural language interpretation confusion and phonetic distance between natural language interpretations was estimated. For measuring the phonetic distance between interpretations, we take all utterances in the test data that lead to a certain interpretation (e. g. "target-aufträge" in Table 4.4) and compute the phonetic distance to all utterances that lead to another interpretation. We take the mean or maximum values as a measure of the confusability between interpretations.

There are correlations between word confusions and (2) articulatory and (3) perceptual phonetic distances. This allows for the introduction of a threshold for confusability prediction. No such correlation was found for the distances between interpretations and the confusion of interpretations.

This can be however due to the small amount of confusion data for inter-pretations. From these results we concluded that we have to base our analysis tool on word similarity. To relate word similarity to interpreta-tions we search for similar words that lead to different interpretations.

4.4.4 Relationship Between User Judgments and Parameters

A method to automatically predict usability judgments is provided by the PARADISE framework [18], in which a linear regression (LR) function is trained on interaction parameters with subjective judgments as targets. PARADISE has mostly been used with parameters describing the dialog flow. The parameters defined in this study extend the conventional set of predictors of user judgments in two ways:

1. Inclusion of signal-level information which potentially contains infor-mation of the users' emotional state.
2. Inclusion of symbolic-level parameters which describe the system char-acteristics; this allows predictions to be made already before the system has been tested in an interaction experiment.

The precondition to predictability of user judgments from parameters is a relationship between the parameters and the user judgments. The training of an LR function, as in PARADISE, further assumes the predictors and target to be correlated linearly. However, in case of known non-linear cor-relations, the parameters can be transformed or alternative prediction func-tions can be used. Therefore, Spearman's ρ, which is sensitive to correla-tions on an ordinal level, was used to examine the potential of our parameters for user judgment predictions.

Concerning the prediction of "overall impression", the strongest, how-ever still moderate relations were found for the channel-related signal pa-rameters (noise level, MOS, MCD) including ASL ($|\rho| \in [0.25;0.31]$, $p<0.01$), and for barge-ins ($\rho = 0.28$, $p<0.01$), followed by the recognizer labels ($|\rho| \in [0.14;0.22]$, $p<0.05$). This means that the impact of channel-related parameters on the judgment cannot be due to their impact on the ASR performance alone, because the latter correlation is lower. An alterna-tive explanation would be that the noise in the user utterance is indicative of the noise the user perceives in the system prompts.

Parameters describing confusability on the word level correlate only weakly with "overall impression" ($|\rho| \in [0.14;0.16]$, $p<0.05$). Thus, this item cannot be predicted from the system characteristics – as long as these are instantiated by the grammar confusability parameters alone – before any interaction takes place.

For "perceived system understanding", highest correlations were found with dialogue-related parameters and recognizer status labels ($|\rho| \in [0.22; 0.46]$, $p<0.01$). Furthermore, phonetic similarities across concepts available in the system grammar are weakly correlated with the same item ($|\rho| \in [0.17; 0.25]$, $p<0.01$). Thus, they explain some crucial part of the recognition problems, as different extents of phonetic similarity are (indirectly) perceivable in the dialog. Ironically, signal-related parameters show no significant correlations with "perceived system understanding" except the noise level ($\rho = 0.22$, $p<0.01$). This is surprising, since they were found to be correlated on high significance level with most of the ASR labels ($|\rho| \in [0.22; 0.40]$, $p<0.01$).

An LR model was trained on the z-transformed parameters (normalizing the input parameters to zero mean and unity variance, and including the parameters stepwise), predicting "perceived system understanding" with an accuracy of $R^2_{adj} = 0.354$. Parameters included in the model are the number of correctly recognized utterances (with a coefficient of -0.399), the number of turns (0.277), the noise level (0.202), and the number of barge-ins (0.147). Training the function on "overall impression" leads to the number of correctly recognized utterances (-0.277), ASL (-0.440), and the number of ASR rejections (0.217) as predictors, describing the data with $R^2_{adj} = 0.225$ accuracy.

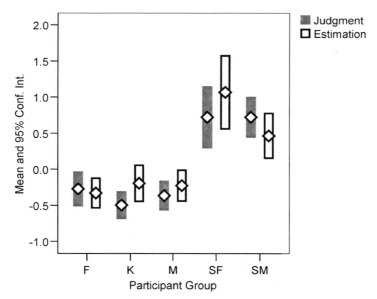

Fig. 4.2. Subjective ratings (filled bars) and PARADISE predictions (open bars) of "perceived system understanding" (mean and 95% confidence intervals) for different user groups (F = females, K = children, M = males, SF = senior females, SM = senior males)

Figure 4.2 shows the mean judgments and PARADISE model predictions for "perceived system understanding", separated by user groups. The mean values for each group can be predicted much better than the values of R^2_{adj} suggest (cf. [19]); especially the bad average judgment of senior users can be predicted by the model. However, the model does not explain why senior males judged as badly as senior females do. Also, the children's mean perception of the system understanding was better than for adults, while the model estimates it to be worse. This indicates that for different user groups different aspects of the system and its usage formed their judgment. Therefore, a more homogeneous sample of users might also have yielded higher correlations in the database.

4.5   Testbed Integration

The algorithms are used in a grammar and dialogue analysis tool with a graphical user interface. This tool is helpful for two different types of analyses. (1) The first type is related to the computation of phonetic similarities between and within grammars and vocabularies. (2) Another type of analysis on the dialogue level can also be performed with the grammar and dialogue analysis tool.

In an analysis that belongs to the first type (1), phonetic similarities between vocabularies for different semantic interpretations in a grammar can be computed. The tab in the graphical user interface on Fig. 4.3 shows

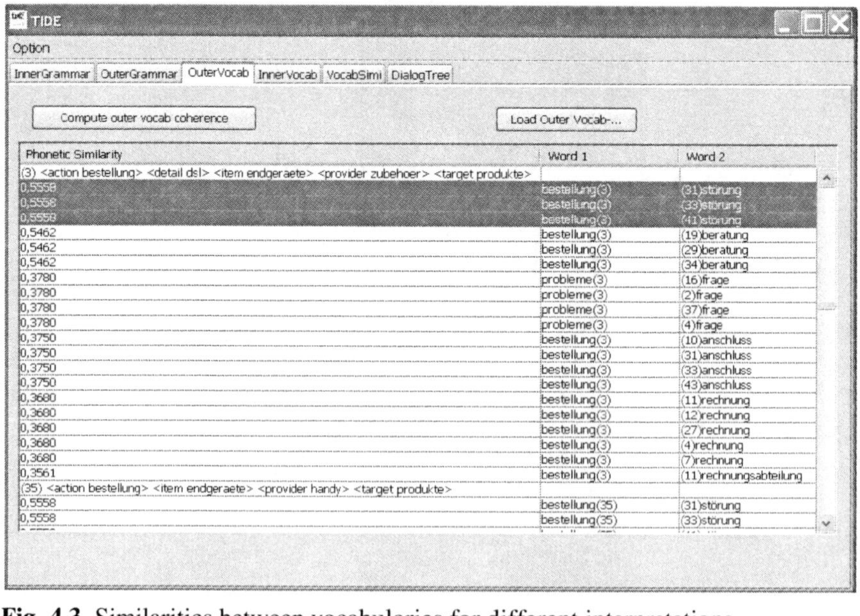

Fig. 4.3. Similarities between vocabularies for different interpretations

words that are phonetically similar and lead to different interpretations. The "Word 1" column contains words that appear in a certain interpretation. "Word 2" shows different words in other interpretations that are phonetically similar to "Word 1". The grammar designer can analyze the similarities and find highly confusable words that lead to different interpretations.

In our test spoken dialogue system, the confusability between "bestellung" and "störung" produced 2% of the understanding errors. As one can see from Fig. 4.3, it is easy to localize this confusability with the grammar analysis tool. Concerning interpretation 3 the three word tokens that are most similar to "bestellung" are all instances of the word type "störung" that appear in different interpretations. When such a high confusability is found, the grammar designer has to decide what to do.

The functions hidden under the other tabs allow highly confusable phrases across interpretations and confusable words and phrases within an interpretation to be found.

Figure 4.4 shows the user interface for the second type of analysis (2) that allows a user to browse dialogues and find problematic dialogue turns, based on "RIPPER" rules derived from signal level parameters.

Dialogue states are extracted from log files of the spoken dialogue system. These states have properties that are derived from the log files or are

Fig. 4.4. TIDE dialogue browser

computed from the signal level. Additionally, properties for the whole dialogue like the number of "rejected" states and the noise level can be loaded.

Problematic dialogue states and dialogues are highlighted with gray arrows, as can be seen in Fig. 4.4. These states are classified through "RIPPER" rules derived from signal level parameters. It is also possible to manually specify dialogue state properties that are regarded as problematic. States having these properties are also highlighted.

Successful dialogue states are dialogue states that lead to an agent connection and are also highlighted. This property can also be set in the option menu. Especially interesting are problematic dialogues that lead to an agent connection. This type of dialogues can be easily detected with this tool. For further analysis it is also possible to listen to the recorded speech, comprising the system prompts and the user input.

These are the two major types of analyses that can be performed with the grammar and dialogue analysis tool. With the help of these analyses, the grammar designer can quickly evaluate and optimize the grammars and dialogues of a spoken dialogue system.

4.6   Discussion and Conclusions

In this chapter, we have addressed the question of how to estimate user-perceived quality and usability on the basis of system and interaction characteristics. For this purpose, we built a testbed which automatically extracts information from vocabulary and grammars as well as from log-files generated with state-of-the-art dialogue platforms. The aim was to optimize dialogue systems with respect to the user's (estimated) perception, and not with respect to individual system performance metrics. Thus, our testbed helps to consequently take a user-centered view-point in the system design process.

On the signal level, parameters have been extracted which allow the sources of recognition errors to be allocated to channel and user characteristics. Active speech duration, the mean cepstral deviation, speech and noise levels, the signal-to-noise ratio, as well as the mean fundamental frequency were shown to be useful for usability prediction. Because the available signals have been collected after the voice activity detection of the speech platform, problems of an improper voice activity detection algorithm cannot be evaluated with our database.

On the symbolic level, phonemic similarities have been computed which allow the confusability of lexicon and grammar to be evaluated. Two phonetic distance measures have been optimized to show an ordering of lexicon items which reflects recognition confusability, and applied to our dialogue database. The results show that an optimized phonetic distance measure can be used for predicting the word confusion and thereby the word error rate, by introducing a threshold. Finally, different coherence measures have been defined which can be used for estimating within- and between-interpretation coherence on a sentence or word level.

We further analyzed the applicability of the extracted information for a semi-automatic usability evaluation, based on predicting recognition performance and user judgments. For this purpose, PARADISE-style linear regression models have been calculated, taking the user judgment on perceived system understanding or the overall impression as the target. Perceived system understanding could be predicted with higher accuracy than overall impression, while neither judgment could be predicted from the parameters extracted with the accuracy achievable with large interaction parameter sets like the ones defined in ITU-T Suppl. 24 to P-Series Recommendations. Correlations between the extracted parameters and recognizer performance have been calculated, showing that neither the signal parameters nor the similarity parameters allow very accurate prediction of recognizer performance. However, it was possible to derive some relations which might be useful when considered in the system design.

The complexity of the task of predicting quality judgments from log data became clear in the light of results of this study. In general, correlations between the extracted parameters and user judgments are very low. E. g., correlation between recognizer performance and the perceived system understanding, though among the highest found in this study, are still below $\rho = 0.5$, which is very low considering that rating and measure in principle describe the same issue – one as a subjective measure and one as an instrumental score. Although signal parameters are correlated with recognizer performance, they do not correlate with the judgment, which shows that concluding from such relations to "what matters" is not a straightforward task.

For the experience of quality of spoken dialogue systems the recognizer performance is of vital importance. Unfortunately, complex, mixed initiative, natural-language-understanding, interruptible spoken dialogues with large and complex grammars make the analysis of ambiguity and confusion the most important issue. Our testbed provides methods that are effective for analyzing and predicting such ambiguities. For example, the fact that the same word can have several meanings within one grammar, or within different grammars running at the same time, usually has serious consequences. Those homonyms have to be tagged with different semantic meanings, depending on the spoken context. Even more problematic are phonetic similarities between different keywords. They are hard to identify but can result in a collapse of the whole application. In short, developers and designers of dialogue systems should be aware of the ambiguities of the implemented vocabulary and grammars.

It is therefore essential to have a tool which alerts them and keeps track of all the potential problems before launching a system. The testbed described in this chapter enables the designer and developer to automatically detect potential problems. In this way, nerve-racking and time-consuming test and error-search periods may be avoided, development costs be reduced, and time-to-market significantly be shortened. It is important to note that such savings coincide with an optimization of the ultimate criterion for spoken dialogue interactions: that they satisfy the user. Thus, semi-automatic usability evaluation with the help of the testbed will finally increase the user acceptance of future spoken dialogue services.

Acknowledgements

The described work was supported by the TIDE project funded by Deutsche Telekom AG. The authors would like to thank all colleagues who supported the run of the experiment and data annotation. The authors gratefully acknowledge Siemens AG, Corporate Technology, Competence Center "User Interface Design", for the fruitful co-operation.

References

[1] Jekosch, U., Voice and Speech Quality Perception. Assessment and Evaluation, Springer, Berlin, 2005.
[2] Möller, S., Engelbrecht, K.-P., Pucher, M., Fröhlich, P., Huo, L., Heute, U., Oberle, F., "TIDE: A Testbed for Interactive Spoken Dialogue System Evaluation", in: Proc. 12th Int. Conf. Speech and Computer (SPECOM'2007), Moskow, Oct. 15–18, 2007.
[3] Pucher, M., Türk, A., Ajmera, J., Fecher, N., "Phonetic Distance Measures for Speech Recognition Vocabulary and Grammar Optimization", in: Proc. 3rd Congress of the Alps Adria Acoustics Association, Graz, Sept. 27–28, 2007.
[4] Möller, S., Pucher, M., Fröhlich, P., Egger, S., Biedermann, J., Türk, A., Ajmera, J., Fecher, N., Engelbrecht, K.-P., Huo, L., Heute, U., "Final Project Report", Deliverable 2, SR Project TIDE (Testbed for Interactive Dialogue system Evaluation), Deutsche Telekom Laboratories, Berlin, 2007.
[5] ITU-T Suppl. 24 to P-Series Rec., Parameters Describing the Interaction with Spoken Dialogue Systems, International Telecommunication Union, Geneva, 2005.
[6] ITU-T Rec. P.851, Subjective Quality Evaluation of Telephone Services Based on Spoken Dialogue Systems, International Telecommunication Union, Geneva, 2003.

[7] ITU-T Rec. P.56, Objective Measurement of Active Speech Level, International Telecommunication Union, Geneva, 1993.

[8] ETSI ETS 300 040, European Digital Cellular Telecommunications System (Phase 1); Voice Activity Detection (GSM 06.32), European Telecommunications Standards Institute, Sophia Antipolis, 1992.

[9] Martin, R., "Noise Power Spectral Density Estimation Based on Optimal Smoothing and Minimum Statistics", IEEE Trans. Speech and Audio Process. 9(5), 504–512, 2001.

[10]Falk, T.H., and Chan, W.-Y., "Single-Ended Speech Quality Measurement Using Machine Learning Methods", IEEE Trans. Audio Speech Language Process. 14(6):1935–1947, 2000.

[11]Möller, S., Quality of Telephone-Based Spoken Dialogue Systems, Springer, New York NY, 2005.

[12]ITU-T Rec. P.563, Single-ended Method for Objective Speech Quality Assessment in Narrow-band Telephony Applications, International Telecommunication Union, Geneva, 2004.

[13]Möller, S., Heimansberg, J., "Estimation of TTS Quality in Telephone Environments Using a Reference-free Quality Prediction Model", in: Proc. 2nd ISCA/DEGA Tutorial and Research Workshop on Perceptual Quality of Systems, Berlin, 56–60, 2006.

[14]Hirschberg, J., Litman, D., and Swerts, M., "Prosodic and Other Cues to Speech Recognition Failures", Speech Communication 43:155–175, 2004.

[15]Rabiner, L., "On the Use of Autocorrelation Analysis for Pitch Detection", IEEE Trans. Acoustics, Speech and Signal Process. 25(1):24–33, 1977.

[16]NIST Speech Recognition Scoring Toolkit (SCTK) Version 2.2.1, National Institute of Standards and Technology, Gaithersburg MD, http://www.nist.gov/speech/ tools/index.htm.

[17]Cohen, W., "Learning Trees and Rules with Set-valued Features", 13th Conference of the American Association of Artificial Intelligence (AAAI), Portland, 709–716, 1996.

[18]Walker, M.A., Litman, D.J., Kamm, C.A. and Abella, A., "PARADISE: A Framework for Evaluating Spoken Dialogue Agents", in: Proc. ACL/EACL 35th Meeting, Madrid, 271–280, 1997.

[19]Engelbrecht, K.-P., and Möller, S., "Pragmatic Usage of Linear Regression Models for the Prediction of User Judgments", in: Proc. 8th SIGdial Workshop on Discourse and Dialogue, Antwerp, 2007.

Chapter 5

Wiebke is the first in line of our authors who share the view that in future systems, while speech may be a prominent mode of interaction, it will not be the only one. We should be aware that any multimodal system raises many more issues than just the sum of the issues relating to each single modality.

As an introduction to the field, Wiebke starts with a practitioner's view on speech and text input and output – using a common example from the telecommunication world.

Dipl.-Ing. Wiebke Johannsen

Wiebke Johannsen qualified as a sound technician in Nuremberg (Schule für Rundfunktechnik) and worked in broadcasting in Berlin for two years. Afterwards she studied electrical engineering at the Technical University in Berlin with emphasis on acoustics and communication science. In 1997 she joined Deutsche Telekom Berkom, which later has become part of T-Systems, as a research scientist in the Speech Quality Assessment group with main activities on the design and evaluation of subjective listening testing in internal projects as well as international standardization bodies. Later, her working field changed from subjective listening tests to usability and acceptance testing mainly in the setting of research projects. Today, Wiebke designs speech applications and multimodal applications focusing on user-specific and personalization aspects.

5 Stylus Meets Voice – a Practitioner's View on Multimodal Input and Output

Wiebke Johannsen
T-Systems Enterprise Services GmbH, Berlin, Germany

5.1 Preface

In the past few years, man-machine-interfaces developed quickly. While the fast increase of internet usage lead to a great demand for good graphical user interfaces (gui) other user interfaces like for example voice user interfaces (vui) profit mainly from the growth of technical possibilities. But still there are lines the interfaces cannot cross by themselves. Considering the natural way humans communicate with each other, it can be observed that – speaking in technical terms – human communication is mainly based on using several input channels simultaneously (e. g. a person says: "give me that" while pointing at a book). Most of the popular man-machine interfaces are only able to use one input channel at a time (e. g. conventionally designed web pages). Therefore most of these man-machine interfaces will be considered to be complicated and unnatural.

The basic idea of multimodal interfaces is to support the human way of communication, to develop applications that everyone will be able to use intuitively without reading long and complicated manuals and without long learning periods before using them properly. The main focus is to offer more than one input channel to the user and to offer them at the same time, so that the user can switch from one channel to another without thinking about it and without even noticing it – just like in a normal dialogue. Of course this concept will also affect the output channels (e. g. providing additional graphic output on the display when asked for railroad connections via voice).

In this article basic thoughts about the development of multimodal inter-faces will be presented. This includes the characteristics of the different input and output channels as well as the basic rules needed to create a mul-timodal interface, which is really a step closer to human communication than the well known one-channel interfaces.

This article will focus on the combination of graphical and voice user interfaces since it is hardly possible to discuss the full spectrum in the con-text of this publication. Also graphical and voice user interfaces are the most common ones at the present time and are therefore well suitable for illustration.

5.2   Basics of Multimodal Applications

5.2.1   Components of Multimodal Applications

The first step to design a multimodal application is to give an overview of the possible input and output channels. Already existing gui- or vui-appli-cations could either be combined or expanded or already existing systems could be re-designed to use the advantages of multimodal systems to full capacity. Combinations of voice and graphics are not the only imaginable and useful ones, more input and output channels can be combined to sup-port special applications. This chapter is intended to give food for thought by introducing commonly known input and output channels.

5.2.1.1 Input and Output Mode

In the beginning of human-machine interaction, communication was strict-ly restricted to one type of input channel only. The user was limited to a key-board in the widest range of meaning since special keyboards could be produced for special input demands (e. g. the telephone keys), but still in-put was restricted to keys. Hardly any other technical way to realize an interface existed. When graphical interfaces and the usage of a computer mouse were introduced, the possibilities almost exploded. Speech based systems also took influence on the variety of possibilities. Each of these systems is suitable for a certain task depending on which input and output channel is used. The following input modes can be used:

- Text: texts can either be typed via keyboard or softkeys or via handwriting recognition
- Click/touch: all input modes in this category perform in the same way in principle. With the help of a graphical user interface symbols or buttons can be chosen simply by clicking them. The most popular devices are:
  - mouse
  - stylus (e. g. PDA in combination with a touchscreen)
  - touchscreen
- Speech: the user can control the system by using speech. There are different ways of speech recognition: single word or keyword spotting and natural language understanding (NLU). Of course, using NLU is more intuitive than using single word spotting, but as technical speech recognition methods are not the focus of this article, they will not be discussed here.
- Gesture/haptics: here everything concerning mimics, body speech and tactual sensations, like pointing at things, the posture, looking at things or movements done with the device itself, are taken into consideration.

As an output mode two different modes are common these days:

- Graphic: this includes all information that humans visually perceive, e. g. videos, photographs, but also text files or tables. Needless to say that the graphic interface is the most important and the need for a professional design is no longer questioned.
- Acoustic signals/speech: this includes all acoustically perceived information. It is mainly speech which transports information. The output is presented with pre-processed speech files or with the use of text to speech systems. In addition, acoustic signals (e. g. earcons or jingles) can be employed. They give the user a hint that something has happened. Feedback of this kind is important. Otherwise the user may not recognize that e. g. he had made a mistake or that a certain event occurs. As soon as one gets used to the application, these signals can support the user.

First steps are being taken to integrate also haptic feedback (e. g. using the vibration alert to confirm input or to signal errors). Since the main focus of this article is set on voice and stylus, no further attention will be turned to this feature.

Not all different input and output modes can be used in every kind of situation. For example an output via speech does not make sense if one is in a noisy environment. Sometimes the user has to do other things at the same time which occupy most of his concentration, so that he is not able to use all input or output channels. A car driver, for example, cannot use graphical interfaces while driving or cannot use his hands for operation, but he can use a speech interface to communicate (e. g. to get to know the latest traffic news). Sometimes the user does not want to work with a certain input channel for security reasons. The user certainly will take care that data like passwords cannot be overheard by another person: Nobody would give his or her PIN to a speech recognition system while traveling in a crowded underground train.

In the following table some typical situations are listed. In this context "partly suitable" means that generally this combination is not suitable but, nevertheless, it is possible to design special applications, which could make sense.

Table 5.1. Suitability of input and output modes

	Input (user)				Output (system)		
	Hands occupied	No attention to display	Noisy environ-ment	Security aspects (bug-ging)	No attention to display	Noisy environ-ment	Security aspects (bugging)
text	not suitable	not suitable	**suitable**	**suitable**	no output mode ***	no output mode ***	no output mode ***
graphic	no input mode	no input mode	no input mode	no input mode	not suitable	**suitable**	**suitable**
click/ touch	not suitable	not suitable	**suitable**	**suitable**	no output mode	no output mode	no output mode
speech	**suitable**	**suitable**	not suitable	not suitable	**suitable**	not suitable	partly suit-able**
gesture/ haptic	partly suitable	partly suitable*	**suitable**	partly suitable*	poten-tially suitable	poten-tially suitable	not suitable

* = suitable if the design is adapted
** = use of headphone necessary
*** = variety of output mode "graphic"

5.2.1.1.1 Intuitive User Guidance

In addition to offer more flexibility, multimodal applications are also one step closer to a natural man-machine communication. One big difference between humans and machines is the way of how information is processed. Machines, at least the ones that are common nowadays work strictly serial: One command after another, in a well defined order. Humans are able to process information in parallel (e. g. they can listen and watch simultane-ously and put the information automatically into the right context). This is also visible in the way how humans communicate with each other. People look at each other while talking to each other. If one of them turns his head around (e. g. because of a noise), the other one will stop talking because he sees (and therefore knows) that his dialogue partner is distracted for a moment. A common voice user interface, for example, would not notice this. When working with this type of machines, this needs to be taken into consideration, or in technical terms: One has to adapt to the system.

For a natural and intuitive usage of man-machine-interfaces the process has to turn: machines should adapt to human habits of communication. Or putting it in a different way: machines have to support multimodal inputs and outputs.

Of course there is no complete solution for machines to distinguish between conscious and unconscious input like humans do, as the machine cannot observe the intention of the user [5]. It is reasonable to question if this aim can be achieved in general, nevertheless first steps in this direction are made (e. g. push-to-talk activation, on-view/off-view detection).

Another aspect is that applications and devices are getting more and more complex and therefore also their usage is getting more and more difficult. Some of the processes cannot be separated into clearly distinguishable units any longer.

5.2.1.1.2 Co-operation of Modalities

A multimodal system should be able to support at least some of the introduced input and output modes. There are different ways of how the different modalities can co-operate:

Two input modes co-operate when they have similar input possibilities. For example: the graphic interface supplies a button named "exit". Saying "exit" should lead to the same result in this case. In addition, the different modes can work together in a coordinated or a non-coordinated way [2]. If the modes work non-coordinated, no exchange of information will occur between the single modes. Each step of the dialogue can be done by each mode, but before each step the user has to choose one input channel. If the modalities work coordinated, all information from all input channels will be put together and will be interpreted at the same time to find out the intention of the user. For example: a user points at a map with a stylus and asks, "Is there a bakery close to this area?" at the same time.

The focus in the technical development is on the coordinated multimodality because there are still a lot of technical questions that need to be solved, like, among others, the problem to guarantee a time-synchronous handling of the input modes and a concept for a joint analysis of the different modes.

5.2.1.2 Devices

The devices have changed a lot in the past few years. Based on the "classic" telephone and PC, a wide variety of portable versions of these two devices has been developed. Also, they get smaller and smaller and get combined with each other (e. g. smart phones).

In principle, there are three different types of devices when talking about telephones and PCs:

- telephones or mobile phones
- PDAs or smart phones
- PCs or Notebooks

The following table shows, which input and output modes are supported by the devices. The focus here lies rather on the useful hardware components and the size of the display than on the computing power. Since this document concentrates on speech and stylus only the relevant modes for these input and output channels are listed.

Table 5.2. Devices – input and output modes

	input (user)			output (system)	
	text	click/touch	speech	graphic**	speech
telephone/ mobile phone	partly suitable (SMS)	not suitable	**suitable**	**suitable***	**suitable**
PDA/ smart phone	**suitable**	**suitable**	**suitable** (need of microphone)	**suitable***	**suitable**
PC/ notebook	**suitable**	**suitable**	**suitable** (need of microphone)	**suitable**	**suitable**

* = restricted because of small display
** = includes text

The progress tends to increasingly compact and mobile devices. But mobility needs small devices, which in turn means restriction in functionality and convenience of use. The small displays of e. g. PDAs can only show little information at a time. Nevertheless, the importance of mobile devices will be growing.

5.3 Designing Multimodal Applications

The creation of a multimodal application does not only include the realization of technical requirements. Another almost even more important aspect is the design of the man-machine interface or better: the man-machine interaction.

The design of a user-focused interface has to support a lot of different, sometimes even opposite, aspects: the technical requirements, the characteristics of the different input and output channels and the habits and abilities of the people using the device. For example, when designing a speech application one should take into consideration that there is a limit in the listing of elements. Due to the capability of the human short time memory, no more than 5 entries should be presented [4]. When designing a graphical user-interface, this is not an important point to take care of.

The goal of multimodal applications is to unite two or more input and output modes to improve the simplicity of use. The number of aspects which have to be taken into consideration will grow with each mode added. The fact that the different modes will interact with each other will even increase this number. The list of facets that have to be taken into consideration will therefore be longer than the one when looking at each single mode separately, which shows the main point: to create a harmonic interface you have to look at the application as a whole.

In the following chapters the four major aspects are introduced which are important in order to design a multimodal interface [3] [6].

5.3.1 Consideration of External Circumstances

The major mantra of creating applications should be, "Help the user to reach his aim as fast and as easy as possible". But what is the fastest way? This mainly depends on the choice of the input and output channel because not all modes are suitable for all surrounding situations. Even the situation itself has an impact on the way the application will work and/or present the results. Not to forget the user: there is a big difference between young, elderly or even disabled people. Depending on their technical background and physical capabilities different designs of the interface might be expedient. In principle there are three categories, which have to be considered:

- the characteristics of the input or output channels that are available
- the personal settings of the user
- the current environmental conditions

There is no mode which is suitable for all surroundings or requirements and sometimes the type of input and output channel has an influence on the efficiency of the mode. Speech for instance is good to enter words and text, but shows its limitation when the name of a certain object is unknown or when a symbol has to be described. On the other hand, it is easy to draw a symbol or even a rough sketch of more complex things, like maps, with a stylus. You can provide pictures of objects which then can be chosen simply by clicking on them (which helps if you do not know the correct name). But for phrases or long text passages speech will still be to best input mode. Conclusion:

> Always favor the input or output mode that is most suitable for the relevant purpose and target group.

This rule cannot be taken in consideration in all possible situations. Imagine a user, who has only limited possibilities to communicate with the system, because he is e. g. just driving a car. He will not be able to do inputs via stylus or concentrate on a complicated graphical output. One has to use other ways in this situation, e. g. speech. There are three typical situations [1] describing the amount of concentration the user can afford for the application:

- "home/office": the user is in a stationary situation, this means he will not change his location and he can completely concentrate on the device.
- "passive mobile": the user is e. g. in an underground train and can concentrate – with few restrictions – on the use of the device.
- "active mobile": the user changes his location by his own (e. g. driving car, walking) and is therefore limited in concentrating on the device.

The application has to be designed in a way, that an easy input or output has to be guaranteed in all possible situations. It has to be planned carefully, in which situation the user could possibly be, when he is going to use the application. Conclusion:

> Pre-select the input and output mode, which offers the best performance in the current situation.

Additional to the limitations that occur because of the mobility of the user, there are also restrictions caused by the current surroundings. It is not sensible, for example, to use speech as an input mode in an environment with a lot of background noise. A general rule says: you can only work with speech as an input signal as long as a normal conversation would be possible as well. On the other hand: if the light is low and there is no way to change the brightness of the display, a graphical output is of no use. Conclusion:

> Adapt the modes to the current environmental conditions.

Of course, even the most talented designer cannot predict how the user will really act and which input or output mode he will prefer and use. The need to support more than one mode at the same time will always be present. But the designer can take into consideration how the majority of users might react in a certain situation. And he should never forget that his own applications might change the environmental conditions, maybe even in a way that the user has to switch the input or output mode (e. g. turn on/off the lights).

5.3.2 Clear Communication on all Levels

One secret of a successful application is to support the user in a way that he will reach his goal without effort. The structure has to be so simple and clear that the user does not even need to think about what he has to do next. This means the application itself should not attract any attention to itself. The user should be able to fully concentrate on getting his business done. He should not be forced to put any effort into the handling of the system. This can be achieved by using clear communication and an interface that is adapted to the habits of the user.

But what is clear communication? An essential step is to use identical cues and prompts throughout all modes. For example: All words that are shown on the buttons of a graphical interface should also be used as keywords for the speech commands. Of course, the system has to react in the same way regardless which mode was chosen. If e. g. there is a button titled "Exit" to quit the application, the word "exit" used as a speech command should also quit the application. There are no limits to expand the amount of keywords for the speech application (e. g. "Quit"), as long as "Exit" remains as a keyword.

The need for more than one keyword in speech applications shows the different characteristics of different input modes. The graphical interface must rely on the recognition of vocabulary by the user. Everyone understands the meaning of the word "exit" when he reads or hears it, but not everyone would use this word to quit an application if he has a free choice. Speech applications deal with the active vocabulary of the user and therefore have to integrate more than one keyword to provide a good support.

With respect to the original topic there is one final thing to mention: all visible buttons should have at least one corresponding speech command. Conclusion:

Use the keywords throughout all modes for identical commands.

Another aspect is consistency. This means that the input or output mode chosen by the user should remain the same throughout the whole communication unless the user changes it. In general the user expects that the system answers in the same way he communicated. When using speech as an input, users usually expect an audio output. Depending on the context, some exceptions are possible. It is pretty clear that a speech command like "Show me the map of Berlin" implies a graphical output. Another example, in which a different output mode is the more appropriate way to answer: when the user asks for "the fastest way to the train station", he may expect an audio output, but a graphical output would support it. As illustrated above, it is possible to add things which may give additional support or which at least reduce the effect of an unwanted change of modes. Conclusion:

> Avoid the change of modes, except when limitations of the chosen mode or the context imply it.

Apart from using the right output mode, it is also very important how the information is presented. Using the right output mode but the wrong way of presentation can lead to a total disaster. In the worst case the information will not be understood at all. A good example for the importance of well-structured information is the reading of telephone numbers. In the first step a short break between the area code and the local code is very important to provide the structure almost everybody is used to. When presenting the number as bundles of two or maybe three, two more advantages will be obvious: first, the number will be easily identified to be a telephone number and second, it is therefore much easier to memorize. Conclusion:

> Structure the information in a way that is commonly used.

Especially telephone numbers can be presented in different ways. Some people make bundles of two; some enumerate one digit after the other. Some will pronounce 2-digit numbers as single numbers; some say it as one number (e. g. 3–4, "three–four" or "thirty–four"). User-friendly systems are able to adapt to the way the user says the numbers and repeat them in the same way. Apart from a good structuring of information, it is always helpful if the system is able to adapt to the user's preferences.

Besides the presentation of information, the design of the user interface is part of a clear communication strategy. Only if the interface is clearly designed, the user will have a feeling of simplicity when working with the application. To increase the clearness you have to take care that the user never looses the orientation. For graphical interfaces you can handle this by using symbols or icons. By using the same symbols consistently throughout the whole application, the user will soon learn their meaning and in consequence will easily find his way through the application. Feeling safe, he can completely concentrate on working with the application instead of serving the application.

For graphical interfaces it is quite easy to gain a good amount of orientation, because seeing does not need much "active" concentration. People simply do it, all the time, subconsciously. If there is no graphical interface to support the orientation, you need much more effort to help the user, because he has to apply concentration which he would prefer to spend on reaching his aim. In speech application systems you can use acoustic signals to guide the user. You can use them to indicate success or failure of the users' input. You can also use earcons to show him which part of the application he is currently using. But you need to be careful: Too many acoustic signals may distract his attention from his original aim. Nevertheless, acoustic signals can be used to support the guidance of the user. The use of signals as an input confirmation can shorten the exposure time in the dialogue which will lead to a higher acceptance. Conclusion:

> Use graphic or acoustic support to gain orientation in user interfaces.

5.3.2.1 Quick and Effective Error Handling

Error handling – especially the handling of mistakes occurring due to wrong inputs or misinterpretation – and the way help functions are implemented is a large and open field. Books could be easily be filled on this topic, so it cannot be discussed in this article satisfactory. But a few basic principles, which should be taken into consideration when designing the error handling of an application, have to be presented.

Assuming that the error occurs because of a malfunction of an input mode, it is always a good idea to support an alternative input mode and – even better – to offer it to the user immediately. So, a prompt can be played while showing a symbol on the graphical interface. Of course, all security aspects have to be fulfilled (e. g. it is no good idea to use an acoustic output for telling the current account balance without knowing if the user is alone). Despite of this fact, a parallel offer of two or more output modes will have the effect that the user will chose intuitively the most suitable medium. Conclusion:

> Offer at least two modes for error handling.

It is self-evident that one can use more than two modes at once for error handling. Especially for help functions this is essential. For example, a user can listen to a prompt explaining him the usage of an application while on the graphical interface the necessary steps are shown. This increases the chance that the user will remember the process when using the application again and therefore will be successful faster. A variation of this proceeding is to play explaining-prompts while the user is already doing the needed steps (e. g. with a stylus). The advantage is that the user reaches his aim after using the support function. Of course, speech cannot be used as an input channel in this example, because a speech input would interrupt the prompt of the support-function.

For a lot of people information about the device itself is important. There are several factors that can take influence on the performance of the device and therefore are significant for error handling, for example the battery status or the quality of the internet connection. Normally, the manufacturer of the device will take care of these aspects. It is not the focus of the application designer, but it is his responsibility to include all the information about the device that is provided by the manufacturer.

5.3.2.2 Layout of the User Interface

The main influence on the judgment of an application is the user interface. If the design is simple and the application can be used intuitively, some small shortcomings of functionality will simply disappear. But, on the other hand, if the design is complex and not clear at all, the user will decide against it by simply not using the application. Even if the functionality is higher than that of alternative products, the user will never use this application intensely enough to explore these features.

Of course, all the aspects that we have already discussed have impact on the design of the user interface. Therefore, they will not be repeated in this chapter.

The most important factor when designing a user interface is the possibility for the user to suppress some of the modes. For example, it must be possible to switch off the acoustic output while attending a meeting, so that no one will be disturbed. Also, security sensitive or personal data should be presented in a bug-proof way. It should be possible to save different profiles, so that a quick switching is possible. Another step is to personalize applications. Conclusion:

Provide control of the input and output modes to the user.

Humans have one big disadvantage: they forget things. And they forget even more if they have to handle a big amount of information in a short amount of time. It is therefore important to present the information in suitable portions. The average short-term memory can master a maximum of seven information-units. Especially voice user interfaces have to consider this fact. Menus or listings should not contain more than five items, otherwise the user will be overburdened. Multimodal applications should use graphical output to support e. g. the output of timetables. Conclusion:

> Support the short-term memory by using graphical interfaces.

One of the most often mentioned criticisms is that the current state of an application is not visible. Imagine the use of a telephone: After picking up the handset and dialing the number, one will hear a dialing tone. This acoustic signal shows that a connection is established and one has to wait until someone on the other side will take up the handset. If the acoustic signal were missing, you would not be able to distinguish if the connection has succeeded or failed. This simple example illustrates the importance of this aspect because almost everyone takes information like this for granted. Especially when the application has to get data from the system or is doing some internal calculation, it is possible that the input channel will not be working for a few seconds. The user might get the impression that the system has stopped working although only some internal processes are activated. You can easily indicate this by showing some graphical output (e. g. a clock in lots of computer applications). Conclusion:

> Provide transparency concerning the system status.

Symbols can also be used as hints. Dependent on the purpose (e. g. hint for a new feature) this can amplify the announcement effect. There are no limits for this, but you have to take care that these symbols have no effect on the clearness of the interface.

5.3.2.3 Motivation to Use Multimodal Interaction

Multimodal applications mark the first step in the progress to design intuitive user interfaces which really support human communication and interaction. Like all technological innovations, multimodal components are not 'natural' for most people. They have to be introduced and, although this sounds contradictory, their usage has to be learned. Most users already have experienced conventional systems (e. g. graphic only user interfaces) and simply do not know that they can choose their preferred way to communicate with the system. The challenge for the designer of multimodal applications is to introduce the possibilities of multimodal interaction while the user is busy dealing with the application. The familiarization with these new potentials has to be designed in an intuitive way. In the best case, the user does not even realize that he gets introduced to something new.

Tests have shown that especially the simultaneous multimodal input is hard to explain and needs time to be accepted by the users [8]. Other tests have shown that users will not always choose a multimodal input. It depends on the situation [7]. The user interfaces therefore should be designed in a way that it is easy to recognize, when simultaneous multimodal inputs are possible. An animated introduction to the multimodal features is surely a good start to illustrate the multimodal principles, but the right way to use such features can only be learned by practicing. There have to be hints, which lead the user to using simultaneous multimodal input. For instance, graphic symbols can highlight multimodal functions. Of course, the hints themselves also need to be designed according to the rules of an intuitive, multimodal interface. How to realize these hints depends on the characteristics and the intended purpose of your application.

The integration of the hints is a big challenge because most multimodal applications are mobile applications and therefore are strictly limited concerning the size of the graphical interface. The user interface could appear to be overloaded very quickly, which opposes the requirement of a clear and structured interface. You also have to take into account that after a while some users of your application might already adapt to multimodal interaction and could be annoyed by too many hints. A well balanced compromise has to be found.

5.4   Summary and Conclusion

Multimodal applications have a big potential to create really new and user-friendly applications on the one side and make especially mobile devices more powerful on the other side.

For designing multimodal applications different aspects have to be taken into consideration. First of all there is the choice which input and output modes are the most suitable ones for the planned application. This depends mainly on the used devices, the assumed environmental conditions and of course of the target group.

In the next step the application has to be carefully designed. Besides the environmental conditions and the knowledge and abilities of the target group a lot of other facets have to be considered: the guarantee of a clear communication, quick and effective error handling and the layout of the user interface as a whole. If the integration of multimodal functionalities is a new feature of the application you also have to introduce the new modalities carefully.

Unnecessary to say that accompanying usability testing might help you to find out what the user expects of you application and how you can improve your design.

The design of a sophisticated multimodal application, which really takes advantage of this potential, is a time-consuming process that needs not only a lot of technical skills but also sensitivity and experience – but it is worth all the effort.

Acknowledgements

The work described in this chapter was funded by Deutsche Telekom Laboratories. I would like to thank the management of the project field "Intuitive Usability" under Katja Henke for the support during the realization of the related projects.

I would also like to thank Dr. Joachim Stegmann and Thomas Scheerbarth as representatives of the whole project team.

References

[1] Ulrich Heid. *Multimodale Dialogsysteme*. Foliensatz zum Hauptseminar Dialogssysteme. Universität Stuttgart (2004).

[2] Knut Kvale, Narada Dilp Warakagoda, Jan Eikeset Knudsen. *Speech Centric Multimodal Interfaces for Mobile Communication Systems*. Telektronic (2003).

[3] James A. Larson. *Commonsense Guidelines for Developing Multimodal User Interfaces* (2003).

[4] George A. Miller. *The Magical Number Seven, Plus or Minus Two: Some Limits on Our Capacity for Processing Information*. The Psychological Review, Vol. 63 (1956).

[5] Wolfgang Minker, Dirk Bühler, Laila Dybkjær (editors). Spoken Multimodal Human-Computer Dialogue in Mobile Environments, 3–11. Springer (2005).

[6] Sharon Oviatt. *Ten Myths of Multimodal Interaction*. Communications of the ACM. Vol. 42, No. 11 (1999).

[7] Sharon Oviatt, Rachel Coulston, Rebecca Lunsford. *When Do We Interact Multimodally? Cognitive Load and Multimodal Communications Patterns*. ICMI'04 (2004).

[8] John Rugelbak, Kari Hamnes. *Multimodal Interaction – Will Users Tap and Speak Simultaneously?* Telektronic (2003).

Chapter 6

Now we come right to the forefront of multimodal research. Roman gives us an in-depth view of the state of research in the field. On the basis of a broad diversity of scientific studies and industry projects, he is the first to have developed a new process model for designing and evaluating any fully multimodal system.

Dr. Roman Vilimek

Roman is a cognitive psychologist who received his doctoral degree "with highest distinction" from Regensburg University. He is an outstanding expert in the field of multi-modal human-computer interaction and received several awards for his work. Roman works for Siemens Corporate Technology's User Interface Design Center.

6 More Than Words: Designing Multimodal Systems

Roman Vilimek
Siemens AG, Corporate Technology, User Interface Design

6.1 Multimodal User Interfaces

Sometimes you want more: With more than one input modality available in a technical system you can interact more effectively across a large range of tasks. Moreover, you can adapt to environmental variables like noise or low visibility or increase efficiency in multiple task situations by supporting an alternation between differently suited input modes. And sometimes you need more: For instance, in order to avoid cognitive tunneling when focusing attention on a very demanding visual task the presentation of task-relevant information outside of the highly attended area in an additional non-visual modality will increase the probability of event detection. Providing the users with multiple input and output modalities in the interaction with a system can make this system more accessible to diverse user groups, accommodate the users' preferences or increase the (feeling of) control on the human side.

Regarding the steadily increasing literature on multimodal user interfaces, there seems to be a wide-spread consensus on the usefulness of designing multimodally. However, probably caused by the interdisciplinary nature of the field different researchers mean different things when describing the properties of multimodal interfaces (cp. Benoît et al. 2000). Depending on the main research some contributions stress the advantages of multiple input modes whereas especially in the psychological context a huge part of available studies concentrates on the effects of addressing multiple sensory modalities, i. e. multimodal output. To get the most out of a multimodal system from a user-centered point of view a more holistic approach is necessary. The human abilities to perceive information and to issue commands have to be more closely aligned to the interaction with a technical device in a given task environment. A good definition in this respect comes from the European Telecommunications Standards Institute. ETSI EG 202 191 (2003) defines *multimodal* as an "adjective that indicates that at least one of the directions of a two-way communication uses two sensory modalities (vision, touch, hearing, olfaction, speech, gestures, etc.)." In this sense, *multimodality* is a "property of a user interface in which: a) more than one sensory is available for the channel (e. g., output can be visual or auditory); or b) within a channel, a particular piece of information is represented in more than one sensory modality (e. g., the command to open a file can be spoken or typed)." The term sensory is used in a broader sense here, meaning human senses as well as sensory capabilities of a technical system. Nigay and Coutaz (1993) point out that "modal" (in multi*modal*) covers "modality" (type of communication channel used to convey or acquire information) as well as "mode" (state that determines in which way information is interpreted).

Multimodal systems have developed significantly during the last two decades. Especially systems involving speech input exhibit a general trend to more robust and natural interaction (Benoît et al. 2000; Kaber, Wright & Sheik-Nainar, 2006; Oviatt, 2003; Reeves et al. 2004). Bolt's (1980) ground-breaking "Put-That-There" prototype with a combination of voice commands, deictic gestures and a large video screen is generally regarded as the starting point of a new area because of the concept demonstration of multimodal fusion. The integration of multimodal feedback, however, has a longer history in applied research. Auditory warnings have always been part of safety critical systems. Tactile feedback has already been evaluated in the 1960s as an output modality in addition to auditory and visual displays for aircraft warning systems (Van Laer, Galanter & Klein, 1960).

The advantages of multimodal interfaces have been demonstrated in various application scenarios. Nevertheless, the success of multimodal applications is not only challenged by increased costs for additional hardware and complex implementation and testing procedures. Relying too much on intuition in design instead of empirically proven knowledge on the users' interaction patterns seriously limits the applicability of multimodal systems (Oviatt, 1999). Anticipated positive effects cannot occur if the users' expectations are not met or if the design of different interaction modalities is not adequate for the task. Users will not profit from multimodal input if they are not able to use all interaction channels equally well and they will not benefit from output for different sensory modalities if they do not understand its meaning. Thus, the successful design of multimodal solutions requires guidance from cognitive science and depends on an early integration of systematic user tests in the development process.

The next section summarizes briefly some experimental results on speech-based interaction in multimodal environments. Subsequently, the effects of multimodal output on perception will be discussed. Special emphasis will be placed on the design process in the last section by addressing the essential question of a systematic and generalizable approach for user-centered development of multimodal systems. A process model will be proposed which combines usability engineering methods with the relevant development steps of the technical realization of multimodal systems.

6.2   Multimodal Input and Interaction Behavior

Using speech in synergistic combination with other interaction modalities changes the linguistic structure of verbal user input. Oviatt (1997) showed that a fewer number of command words and thus shorter sentences and less complex constructions are the consequences of a fusion of speech and manual or gestural input. Of course the simplifying effect of multimodal input is most prominent in tasks with spatial components. Here, unimodal verbal input is not suited well. Trying to define spatial parameters and locations results in long and cumbersome natural language constructions with a large number of self corrections (Oviatt, 1997; Oviatt & Kuhn, 1998). This is not very surprising in itself, but it is quite astonishing how quickly users can adapt to the characteristic advantages of different input modalities and how this simplifies their language.

The fusion of input modalities requires a well-founded hypothesis about the users' intention. When performing verbal and manual input in short succession: Did the user want to issue *one* multimodal command implying a complementary or a synergistic merging of the modalities? Or should the information from the different channels be interpreted separately in terms of *two independent* commands? Time-stamps and temporal thresholds form in many cases the basis for this decision (Allen, 1983; Nigay & Coutaz, 1993; Oviatt, 2003). Establishing a good criterion is essential as vast differences exist between users and tasks. An additional factor in multimodal constructions is that the integration of for example speech and manual input may be sequential or simultaneous. Whereas in simultaneous integration both input modalities are partially overlapped in time, there is a certain time lag in sequential integration after the end of a command in one modality and the beginning of the other (Oviatt, Lunsford & Coulston, 2005). These intermodal lags can take up to four seconds (Oviatt, DeAngeli & Kuhn, 1997). In order to avoid that systems need to wait this relatively large period of time – which would slow down the interaction process – Oviatt et al. (2005) suggest to make use of stable individual preferences in integration patterns and implement user adaptive instead of fixed temporal thresholds. Sharon Oviatt and her research group (Oviatt et al. 2003; Oviatt et al. 2005) were able to show that about 70% of all users tend to simultaneous integration and that a dominant (simultaneous or sequential) integration pattern can be identified within the first few multimodal commands which remains stable over time. An appropriate classification procedure in multimodal systems can substantially reduce delays for dominantly simultaneous integrators.

As far as usability is concerned, however, sequential integration of speech and manual input seems to be preferable. The long-term study of Oviatt et al. (2005) revealed that sequential integration leads to a far simpler command-like language with a smaller user vocabulary with less variations. This led as a direct consequence to less input errors and fewer self-corrections, pauses, interjections and hesitations. Earlier experiments have already shown that spontaneous spoken disfluencies increase linearly with progressively longer utterances due to higher planning demands (Oviatt, 1994). Considering that especially in dual-task situations the quality of speech production is further reduced (Baber & Noyes, 1996) it may be beneficial to keep systems simple by shaping user behavior to sequential integration. Of course this needs careful evaluations to ensure acceptance – especially if the user has a natural tendency for simultaneous integration.

Whether users will tend to formulate their wishes in natural language or in terse command style also depends on the application area. Experiments with speech-based infotainment and navigation systems for cars show a clear preference for command style over conversational input (Graham, Aldridge, Carter & Lansdown, 1999). In general, many aspects of users' verbal behavior depend on the perceived identity of a system (Richards & Underwood, 1984). As most machines and automated systems do not display human-like qualities, command-based interaction will match peoples' expectations especially for first or casual contacts (Akyol, Libuda & Kraiss, 2001; Richards & Underwood, 1984; Zoltan-Ford, 1991). This fact should be kept in mind when making use of the growing conversational capacities of modern speech dialog systems – a simple but well-designed command-based voice user interface should always be available in a multimodal system as a fallback (Vilimek & Hempel, 2005b; Vilimek, Hempel & Otto, 2007).

Not all modalities are equally suited to convey a specific class of information. Wickens and Hollands (2000, p.419) stress that the human voice is especially adapted to transport linguistic, symbolic and categorical information like "acquire the red square". Analog information in contrast, like "a little to the left; now up a bit" can better be accomplished by manual input. They conclude that speech as an input modality for position control will only make sense if the target destination can be unambiguously determined. As each modality (voice, manual) has an inherent compatibility for its favored information (symbolic, spatial) these coherences should be respected in the assignment of primary modalities to a task.

Some of the advantages which are typically attributed to multimodal input do not necessarily show up in real life. Bengler (2001) reports that the range of available interaction possibilities is hardly used. Furthermore, one of the common assumptions on the combination of probabilistic modalities like speech recognition and conventional manual input is not well proven in practice: In theory, in the case of multiple recognition errors people could flexibly change the input modality to a less error-prone technology. In practice, users were not willing to change a once selected way of input for a certain task step (Althoff & McGlaun, 2001, as cited in Bengler, 2001). Considering operator workload, multiple input modalities do not always reduce strain. If, for instance, in order to use gesture input new interaction principles have to be learned, higher mental and physical demands can be demonstrated even over a long period of time (Keates & Robinson, 1999).

6.3   Multimodal Output and Perception

Systematic research on the effects of multimodal stimuli on human perception started almost one hundred years ago with Todd's (1912) experiments on the redundant signals effect (RSE). Todd presented his subjects with visual stimuli with or without a simultaneous auditory stimulus. The bimodal stimuli led to significantly shorter manual reaction times. The facilitation effect can also be achieved with tactile stimuli, is influenced by stimulus intensity and onset asynchrony and can be further increased by trimodal signals (Diederich & Colonius, 2004). It is quite interesting that despite age-related deficits in divided attention the ability to profit from multimodal stimuli is preserved in advanced age (Bucur, Allen, Sanders, Ruthruff & Murphy, 2005). However, although the RSE is very stable and reproducible, the magnitude of the effect is not sufficient in itself to justify the increased effort of implementing multimodal feedback. In general, the reduction of response times lies between 20 to 90 milliseconds (Gielen, Schmidt & van den Heuvel, 1983). Nevertheless, his basic research finding is highly relevant for interaction design as it points to a specific capacity in human information processing.

For some application scenarios reaction time enhancement is less important than detection probability. Early basic research showed that redundant multimodal presentations can increase the probability of stimulus detection (cp. Buckner & McGrath, 1963; Howarth & Treisman, 1958). Although this finding was replicated in some applied research studies (for a discussion see Wickens & Hollands, 2000), several experiments did not find particular advantages in presenting a stimulus for a single task with redundant output (e. g., Kobus et al. 1986; Seagull, Wickens & Loeb, 2001; Tzelgov, Srebro, Henik & Kushelevsky, 1987). As redundant multimodal output never leads to a lower detection probability but bears at least the chance of better detectability, Kobus et al. (1986) suggested to always prefer it if suited for the task.

Anyway users can only then profit from a redundant multimodal signal if they perceive it as *one* simultaneous event. This depends on the already mentioned stimulus onset asynchrony. If a time lag between an auditory and a visual stimulus exceeds a certain maximum, the signals will be regarded as two consecutive events. General recommendations on the maximum delay range between 25 ms (Altınsoy, 2006) and 50 ms (Card, Moran & Newell, 1986). Especially even very small auditory-tactile time lags are easy to detect. For most modality combinations an asymmetry in sensitivity for the detection threshold in signal lead and signal lag exists (Altınsoy, 2006; Hempel & Altınsoy, 2005; Kohlrausch & van de Par, 2005), making humans comparably more sensitive for audio lead with respect to a visual signal (40 ms audio lead vs. 100 ms visual lead) and for auditory lead with respect to a tactile signal (25 ms auditory lead vs. 50 ms tactile lead). Between visual and tactile stimuli there seems to be no such asymmetry (45 ms in both directions; Vogels, 2004).

Focusing more directly on experiments in the HMI context, where reaction time and detection probability are only two relevant variables among others, benefits of multimodal feedback are well proven. Multimodal feedback has been associated with a lower error rate, shorter task completion times and other task performance-related measures, better time-sharing performance and decreased workload in comparison to unimodal output (e. g., Arsenault & Ware, 2000; Brewster, 1998; Cockburn & Brewster, 2005; Göbel, Luczak, Springer, Hedicke & Rötting, 1995; Kaber et al. 2006; Liu, 2001; Oakley, McGee, Brewster & Gray, 2000; Vilimek & Zimmer, 2007; Vitense, Jacko & Emery, 2003). Especially older users may increase their level of performance with additional feedback (Jacko et al. 2004; Jacko et al. 2005; Liu, 2001).

For some use cases positive effects arise by specializing different modalities for different events like e. g. auditory warnings (Edworthy & Adams, 1996). But when it comes to effects of divided attention which are typical for multiple task situations, great care has to be taken not only in the design of these warnings but also to avoid negative effects of attention switching. Spence and Driver (1997) discuss some of these effects. If a person has to react to an event in one modality but expected a piece of information in another modality (e. g., because events in this modality happen more often), the reaction performance gets significantly slowed down. If one of several sources of information gets more important because of a higher event probability, the attention can get focused on this channel and rare events in another modality may not be answered optimally.

In the context of multiple task situations and time-sharing performance it is very helpful to consider theoretical models. In the late 70s, capacity models introduced the concept of multiple resources in attention and information processing to explain human performance in multiple task situations and capture effects like training and automation (Kantowitz & Knight, 1976; Navon & Gopher, 1979). The more two concurrent tasks rely on independent resources, the weaker the interference between these tasks. Wickens (1984, 2002) proposed an elaborated multiple resource model of human information processing and provided an insight into the nature of these resources. Based upon a meta analysis of a wide variety of multiple task experiments as well as on his own research results he identified several structural dimensions that can be conceptualized as separate resources. In the original version of the model three independent dimensions have been identified: stages of processing, perceptual modalities and processing codes. The modality dimension is of course a vital aspect for the design of multimodal interfaces. As long as all other resource demands are equal, the model states that two tasks which both demand visual perception will interfere with each other more than two tasks that demand different levels of that dimension, i. e. one visual and one auditory task. Benefits of multimodal feedback can be nullified, however, by increased demands in other resource dimensions, for instance in case of a cognitive bottleneck (central processing). Furthermore, it is not only important to take into account to use independent modalities but also to consider the effects of processing codes: If for instance one task involves reading visually presented verbal information and another task consists of shadowing auditory verbal information there will be a large interference between these tasks despite the different presentation modalities. Similarly, auditory output may reduce visual load, but different types of auditory system messages (verbal/nonverbal) have different effects on central resources like the limited capacity of working memory (Hempel & Vilimek, 2007; Vilimek & Hempel, 2005a).

Although it is quite apparent that attention can be divided better between the eye and the ear than between two auditory or two visual sources of information the relative advantage of cross-modal over intramodal time-sharing may not always be the result of using separate perceptual resources but rather of peripheral factors. If for instance two displays which theoretically allow parallel scanning are placed far apart from another, additional costs arise in terms of necessary head or body movements. Spence and Driver discuss various other findings that question the multiple resource model (Driver & Spence, 2004; Spence & Driver, 1999). Nevertheless, the model provides good guidance for the design and evaluation of multimodal interfaces because it helps in asking the right questions.

It is self-evident that in absence of a complete theory which allows deducting the effects of multimodal design on human performance the empirical evaluation of design variables becomes crucial. Considering the complexity of multimodal interfaces user testing and ensuring the usability of a multimodal system is one of the most difficult aspects in design. The process model which is introduced in the next section tries to provide a framework for a systematic approach on this topic.

6.4   Making it Real: the Design Process

One of the most prominent challenges in designing multimodal user interfaces is to find a way for systematic user testing and affordable prototyping efforts despite the huge complexity of these systems. A common strategy is to build a working prototype of a multimodal system first and then to try to evaluate whether the desired effects on human performance occur. Thus, one can assure that all components of the system work properly before engaging in user tests and it is possible to work directly on the target platform. On the other hand, there are three significant disadvantages. First, it is almost impossible to keep the prototype flexible enough for a systematic test of all design variables of interest. The effort to implement this configurability simply exceeds the budget provided for programming. Secondly, even if the first aspect can be handled, putting all relevant questions on variants of input and output into one experiment would require a very large number of test participants in order to get sufficient sample data. Schomaker et al. (1995) point out that the number of experimental conditions grows with each additional independent variable due to combinatorial explosion. The needed number of experimental subjects as well as the required strict experimental control over the design variables can hardly be realized. Thus, a lot of important design decisions in the construction of a multimodal system are typically based on intuition and not explicitly tested. Thirdly: The more complex a prototype is, the harder any changes in hard- or software will get. If the results of the user tests indicate major overthrows, the manpower and costs for these changes may threaten a successful completion of the project.

Therefore, a process needs to be established which allows to address the relevant design questions 1) with as far as possible simple and easy to set up prototypes 2) in a way that allows to decompose the user tests into small separate experiments which can be handled systematically and 3) early enough in the implementation process to keep costs controllable and to allow for a coordinated information exchange between user experience experts and technical experts. These prerequisites form the basis of the proposed design process. By integrating typical steps and problems of the implementation of multimodal systems with an iterative and evolutionary user-centered design approach, a solid foundation for systematic research as well as for successful product solutions can be achieved.

User-centered design means first of all to have a very good understanding of the users. In the context of the first book exclusively focusing on this subject (edited by Norman & Draper, 1986), Norman (1986) states that "the needs of the user should dominate the design of the interface, and the need of the interface should dominate the design of the rest of the system." Thus, the context of use, the users' goals, tasks and needs play a dominant role. There are several key principles for user-centered system design. Besides the already mentioned user focus which implies user involvement at early development stages, some aspects summarized by Gulliksen et al. (2003) deserve special attention. One aspect is evolutionary systems development, i. e. the development process should be iterative and incremental. The system needs to be divided into parts which can be delivered for real use and the evaluation context should reflect relevant characteristics of the target use context. As early as possible, repeated tests with simple prototypes should be carried out. Another aspect is the integration of usability know-how, either by involving usability experts or by narrowing down the range of possible design solutions based on published research results. And, as already mentioned in the first section of this paper, a holistic design approach is essential. Optimizing only one interaction channel without analyzing the whole interaction process might lead to benefits where they are not needed while leaving the real bottleneck in the system unaltered.

The ISO 13407 (1999) defines specific requirements for any user-centered design process. The process framework of ISO 13407 can be viewed as the common foundation of many usability engineering process models which vary in scope and level of detail according to their specific focus (Reichenauer, 2005). According to this standard, four main activities should take place during system development:

- Understand and specify the context of use.
- Specify the user and organizational requirements.
- Produce design solutions.
- Evaluate designs against requirements.

The process model proposed here (see Fig. 6.1) reflects the considerations on the general usability engineering process as described above and

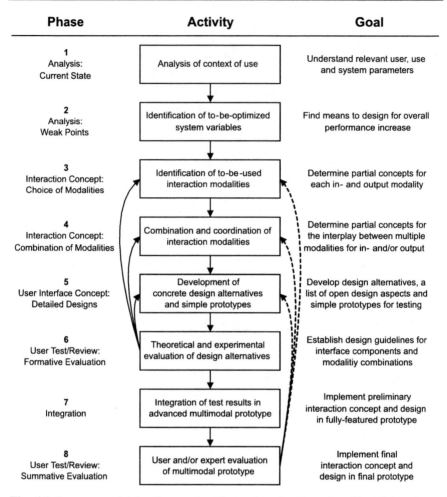

Fig. 6.1. Process model for the construction and evaluation of multimodal systems

merges them with the special requirements of the design of multimodal systems. One of the most important aspects of the model is the breakdown of the empirical evaluation in several sub-activities. Instead of working directly on the final prototype, early work on simple prototypes and the integration of theoretical knowledge on multimodal interaction are explicitly included. Besides the known differentiation between formative and summative evaluation, there is a split of the formative evaluation into a theoretical and an experimental test part addressing the design alternatives for the different interface components, i.e. alternatives of modality-specific and multimodal design. The specification of each phase of development, the description of activities within these phases and the intended results or goals is supposed to contribute to a transparent project or research plan for multimodal systems which increase the probability of delivering the expected advantages of multimodal interaction.

The process model tries to solve some basic problems of multimodal design. Some important steps to be taken in interaction and user interface design in general cannot be elaborated here. For an overall approach it is necessary to combine the aspects of multimodal design as described here with established design methods like e. g. scenario-based design (Rosson & Carroll, 2001, 2003). This method also includes the evolutionary design principle starting with a basic interaction concept and moving forward to a detailed user interface design concept (or from a "root concept" to an "articulated concept" in the words of Go & Carroll, 2004) but is especially suited to avoid the hazards of a "solution-first" strategy. It helps to analyze what is already known about a problem and the possible solving strategies. Thus, it can guide the process of solution generation and help to find situations in which multimodal interaction principles can provide optimal benefit.

In the following, the different phases of the process model will be introduced by posing the central questions of the respective process step. As an illustration, examples from the design of a multimodal system for advanced in-vehicle interfaces will be provided. In order to keep explanations simple these illustrations will not be exhaustive or cover the use case in full detail.

6.4.1   Phase 1: Analysis – Current State

The first phase of the model is focused on the question *How is the given context of use characterized?* The ISO standard 9241-11 (1998) on usability defines that this context comprises of a description of the users, tasks, equipment (in terms of hardware, software, material associated with the task) and relevant environment variables. A good representation of the context of use is not only essential for design; the respective parameters also serve as criteria for the ecological validity of the final summative evaluation. For in-vehicle interaction, only a few characteristics of the users help in narrowing down the target group. All drivers need a driver's license, but the test may lie back in the distant past. The design of on-board systems has to consider a really large user group, containing casual users, changing users (e. g., drivers of rental cars) and non-professional users.

These circumstances pose high demands on the selection of test participants in user tests – at least for the later stages of evaluation. Considering task and environment, the design may be guided by well-proven hierarchical models of driving which describe the demands on the driver associated with different aspects of the task (e. g., Michon, 1985). The interaction with on-board devices increases the workload by introducing a multiple task situation with high visual load. Additionally, the steadily increasing amount of additional functions needs to be handled. To avoid an excessive amount of controls and displays, a growing number of car manufacturers rely on a menu-based interaction concept with a display in the center console and a single central control for manual input (e. g. a rotary push-button).

6.4.2   Phase 2: Analysis – Weak Points

The second phase is based upon the analysis results of the prior step: *Which aspects of the interaction situation need to be improved?* For instance, the large number of comfort functions increases the time needed to locate a specific function thus increasing the gaze durations at the display. Multimodal in- and output may help the driver to allocate more attentional resources back to driving. It is very likely that these weak points are already known a priori or are even the reason for the decision to build a new system. Nevertheless, both the first and the second phase should be walked through with conscious effort. This will increase the chance to identify wrong assumptions on optimization potential early enough.

6.4.3   Phase 3: Interaction Concept – Choice of Modalities

The central question of the third phase can be formulated as *Which elements of the problem space can be improved by multimodal interaction?* Theoretical models like the multiple resource model and experimental as well as other empirical findings from multimodal interaction research can provide a helpful framework for this phase. Finally, one or several partial concepts for the employment of in- and/or output channels need to be established. For instance, realizing additional auditory output or speech input may reduce the visual load of the user while driving. Conducting first tests on technical feasibility of the desired solutions already in this phase will help to detect technical challenges early enough to choose an alternative solution or to find a workaround.

6.4.4   Phase 4: Interaction Concept – Combination
of Modalities

A concept for the combination of the modality candidates will be developed in phase 4 (*How should the available in- and output modalities co-operate?*). In this respect, the results of phase 3 pre-determine this phase. Partial concepts for each combination form are considered in still comparably abstract form. Different suggestions for these combinations each require appropriate interaction concepts as they typically come along with a change in the logic of operation. This becomes clear when regarding the possible types of cooperation in detail. Based upon the terminology of Martin (1997), Benoît and colleagues (2000) differentiate between six types of cooperation which relate primarily to input modalities but are extendible to a certain degree to output modalities as well: complementarity, redundancy, equivalence, specialization, concurrency and transfer. On the output side, one has to ask for instance whether redundant output by equivalent modalities should be preferred or whether each output modality should be uniquely assigned to a special and distinct class of events by specialization. On the input side a decisions has to be made if a synergistic combination of modalities would make sense. In the interaction with a multimodal system this combination is typically more characterized by complementarity than by redundancy (Oviatt, 1999). In many cases, parallel ways of input (equivalence) will be necessary to ease interaction in multiple task situations, e. g. by providing equally powerful manual and voice control.

The holistic design approach is advantageous especially in this phase. Frequently, an additional input modality like speech recognition is introduced on top of an existing interaction concept and the optimal combination of modalities is not considered. The best results will be achieved by integrating the interaction concepts for each modality at this early stage of the development process and by regarding the means of in- and output as a whole, i. e. not designing them in isolation. The concepts in this stage will still leave a lot of freedom for interaction design. In this sense they can be regarded as a basis or framework for the development of concrete design proposals. Crucial at this point is furthermore a very clear picture of the technical feasibility. The architecture and demands on hard- and software of some types of cooperation are more difficult to realize than others. Although a high technical quality is not sufficient for usability, it is definitely a necessary prerequisite.

6.4.5 Phase 5: User Interface Concept – Detailed Designs

The abstract ideas produced in the phases of working on the interaction concept will be transformed to use case and user oriented interface concepts in form of design drafts with the central question *Which concrete design variations are imaginable with the planned in- and output modalities and their planned combinations?* Although phase 3 and 4 provide direct input for this phase, there is still a lot of work to do. Like visual design is one of the most difficult tasks for conventional user interfaces, multimodal systems need sound design, haptic design, voice dialog design, etc.

Each of these subsets alone implies respecting special guidelines. In addition to these unimodal questions, special design requirements exist for multimodal design itself (e. g., Larson, 2006; Reeves et al. 2004).

The plural in "detailed designs" already indicates that typically different design solutions will result. Some of them may complement each other, others may conflict and form completely different alternative solutions. In many cases it will not be possible to make a choice a priori, so some design aspects must be tested with simple prototypes or experimental approaches. For instance, a unimodal design question can be whether to use speech- or non-speech sounds as an additional non-visual feedback modality. Besides influences from the physical and social characteristics of the task environment, this decision depends on characteristics of the sounds or speech messages to be used and may result in questions like "How good can the users distinguish between these sounds?" or "Does verbal output interfere more or less with other task components?", etc. Furthermore, all auditory events need to be designed in detail. For sounds this means among other things to specify presentation technique, frequency, amplitude, loudness, duration, direction, rhythm, dynamics, timbre, pitch and register. The goal of this phase is to get all user interface components to a level on which they can be tested empirically or reviewed theoretically.

6.4.6   Phase 6: User Test/Review – Formative Evaluation

In Phase 6 decisions will be made on the suitability of the available alternative designs. Open design aspects will be clarified by theoretical review or empirical user testing. The efforts should answer the question *Which design variables lead to a successful multimodal UI concept?* and deliver unambiguous design guidelines based upon a thorough review. Some of these guidelines can be achieved by theoretical review, i. e. consulting handbooks and other relevant literature. Other open issues need separate experimental testing. Regarding the example on sound design given above it makes sense to try to find the best design alternative first within the different modalities and conduct experiments on this modality as far as possible in isolation to keep the number of experimental conditions manageable. Subsequently, aspects across modalities need to be addressed. For instance, testing starts with an experiment on voice input and gathers subjective and/or objective data if speech is a useful modality in the given context or which speech input strategy should be supported. The next step may be to evaluate which combination of voice and manual input (equivalent, specialized, etc.) delivers the best results. As this procedure allows to address the relevant questions in separate experiments, simple prototypes are sufficient. The whole complexity of multimodal interaction does not necessarily need to be simulated at this point. Focusing design questions in single experiments leads furthermore to a higher degree of experimental control and reduces artifacts.

Like in any formative evaluation procedure, the results of the experiments or the theoretical reviews can lead to a concept or design rework. This is illustrated by solid arrows in Fig. 6.1 pointing back to the concept phases. Any conceptual changes need of course to be reviewed again.

6.4.7   Phase 7: Integration

Phase 7 concentrates on the problem *Which design alternatives proved to be useful after formative evaluation and may be integrated into a fully-featured prototype?* These interaction and design concepts form the foundation of the architecture and the properties of the multimodal system. In this phase – and not sooner – the work on realizing the multimodal target system starts. Before engaging in final summative user tests the prototype needs careful technical testing to ensure full functionality.

6.4.8   Phase 8: User Test/Review – Summative Evaluation

The last phase serves as a final summative evaluation with usability tests and usability expert reviews. The test setting should resemble the intended use context as far as possible, i. e. typical environmental and task variables need to be simulated and the test participants should represent typical users to answer the question *How do the users interact with the multimodal system in a real-life setting?* This means for example in the context of in-vehicle interaction to use driving simulators or real driving tests on a closed roadway. It is important to understand in which way users interact with the system, how they utilize multimodal interaction principles, whether modality preferences will emerge and especially if there are any problems. Special attention should be given to the question whether the identified successful solutions from unimodal tests (phase 6) are equally successful in the overall test or whether the integration of these solutions into a multimodal system may have caused a completely new situation leading to an altered user behavior. The test results of the prior experiments in phase 6 may serve as a baseline for this decision when inspecting the new data of summative evaluation. In many cases only a few problems will be found at this stage when following the proposed process. However, sometimes a rework of some design aspects may yet be necessary although a summative test is intended to be the final step. The process model depicts this by dashed arrows (instead of solid lines) pointing back to the conceptual phases and allowing for a final iteration step.

6.5 Summary

Multimodal in- and output can increase task performance in the interaction with technical systems. The advantages of multimodal user interfaces will only show up if they are designed to support the abilities and characteristics of the users. Thus, it is necessary to integrate research results from cognitive sciences in the development process. The experimental findings discussed in this paper demonstrate this necessity. The proposed process model for multimodal design shows how the interplay between the development of system components and user-centered evaluation can be organized.

References

Akyol, S., Libuda, L. & Kraiss, K. F. (2001). Multimodale Benutzung adaptiver Kfz-Bordsysteme. In T. Jürgensohn & K. P. Timpe (Eds.), *Kraftfahrzeugführung* (pp. 137–154). Berlin: Springer.

Allen, J. F. (1983). Maintaining knowledge about temporal intervals. *Communications of the ACM, 26*(11), 832–843.

Altinsoy, E. (2006). *Auditory-tactile interaction in virtual environments.* Aachen: Shaker.

Arsenault, R. & Ware, C. (2000). Eye-hand co-ordination with force feedback. In *Proceedings of the SIGCHI conference on human factors in computing systems* (pp. 408–414). New York: ACM Press.

Baber, C. & Noyes, J. (1996). Automatic speech recognition in adverse environments. *Human Factors, 38*(1), 142–155.

Bengler, K. (2001). Aspekte der multimodalen Bedienung und Anzeige im Automobil. In T. Jürgensohn & K. P. Timpe (Eds.), *Kraftfahrzeugführung* (pp. 195–205). Berlin: Springer.

Benoît, J., Martin, C., Pelachaud, C., Schomaker, L. & Suhm, B. (2000). Audiovisual and multimodal speech-based systems. In *Handbook of multimodal and spoken dialogue systems: Resources, terminology and product evaluation* (pp. 102–203). Boston: Kluwer Academic Publishers.

Bolt, R. A. (1980). Put that there: Voice and gesture at the graphics interface. *ACM Computer Graphics, 14*, 262–270.

Brewster, S. A. (1998). The design of sonically-enhanced widgets. *Interacting with Computers, 11*(2), 211–235.

Buckner, D. N. & McGrath, J. J. (Eds.). (1963). *Vigilance: a symposium.* New York: McGraw-Hill.

Bucur, B., Allen, P. A., Sanders, R. E., Ruthruff, E. & Murphy, M. D. (2005). Redundancy gain and coactivation in bimodal detection: Evidence for the preservation of coactive processing in older adults. *Journals of Gerontology: Psychological Sciences and Social Sciences, 60*(5), 279–282.

Card, S. K., Moran, T. P. & Newell, A. (1986). The model human processor: An engineering model of human performance. In K. R. Boff, L. Kaufman & J. P. Thomas (Eds.), *Handbook of perception and human performance* (Vol. 2: Cognitive processes and performance, chap. 45, pp. 1–35). Oxford, UK: John Wiley & Sons.

Cockburn, A. & Brewster, S. A. (2005). Multimodal feedback for the acquisition of small targets. *Ergonomics, 48*(9), 1129–1150.

Diederich, A. & Colonius, H. (2004). Bimodal and trimodal multisensory enhancement: Effects of stimulus onset and intensity on reaction time. *Perception & Psychophysics, 66*(8), 1388–1404.

Driver, J. & Spence, C. (2004). Crossmodal spatial attention: Evidence from human performance. In C. Spence & J. Driver (Eds.), *Crossmodal space and crossmodal attention* (pp. 179–220). Oxford, UK: Oxford University Press.

Edworthy, J. & Adams, A. (1996). *Warning design: A research prospective.* London: Taylor & Francis.

ETSI EG 202 191. (2003). *Human Factors (HF); Multimodal interaction, communication and navigation guidelines.* Sophia-Antipolis Cedex, France: ETSI. Retrieved December 10, 2006, from http://docbox.etsi.org/EC_Files/ EC_Files/ eg_202191v010101p.pdf.

Gielen, S. C., Schmidt, R. A. & van den Heuvel, P. J. (1983). On the nature of intersensory facilitation of reaction time. *Perception & Psychophysics, 34*(2), 161–168.

Go, K. & Carroll, J. M. (2004). Scenario-based task analysis. In D. Diaper & N. A. Stanton (Eds.), *The handbook of task analysis for human-computer interaction* (pp. 117–134). Mahwah, NJ: Lawrence Erlbaum Associates Publishers.

Göbel, M., Luczak, H., Springer, J., Hedicke, V. & Rötting, M. (1995). Tactile feedback applied to computer mice. *International Journal of Human-Computer Interaction, 7*(1), 1–24.

Graham, R., Aldridge, L., Carter, C. & Lansdown, T. C. (1999). The design of in-car speech recognition interfaces for usability and user acceptance. In D. Harris (Ed.), *Engineering psychology and cognitive ergonomics: Job design, product design and human-computer interaction* (Vol. 4, pp. 313–320). Aldershot: Ashgate.

Gulliksen, J., Göransson, B., Boivie, I., Blomkvist, S., Persson, J. & Cajander, A. (2003). Key principles for user-centered systems design. *Behaviour & Information Technology, 22*(6), 397–409.

Hempel, T. & Altınsoy, E. (2005). Multimodal user interfaces: Designing media for the auditory and the tactile channel. In R. W. Proctor & K.-P. L. Vu (Eds.), *Handbook of human factors in web design* (pp. 134–155). Mahwah, NJ: Lawrence Erlbaum Associates.

Hempel, T. & Vilimek, R. (2007). Zum Einfluss von sprachlichen und nicht-sprachlichen Systemausgaben auf Arbeitsgedächtnis, Reaktionszeit und Fehlerrate: Grundlagen für den Einsatz im Kfz. In S.-R. Mehra & P. Leistner (Eds.), *Fortschritte der Akustik – DAGA 2007* (pp. 299–300). Berlin: DEGA.

Howarth, C. I. & Treisman, M. (1958). The effect of warning interval on the electric phosphene and auditory thresholds. *Quarterly Journal of Experimental Psychology, 10*, 130–141.

ISO 9241-11. (1998). *Ergonomic requirements for office work with visual display terminals (VDTs). Part 11: Guidance on usability.* Geneva, Switzerland: International Organization for Standardization.

ISO 13407. (1999). *Human-centered design processes for interactive sytems.* Berlin: Beuth.

Jacko, J. A., Emery, V. K., Edwards, P. J., Ashok, M., Barnard, L., Kongnakorn, T., Moloney, K. P. & Sainfort, F. (2004). The effects of multimodal feedback on older adults' task performance given varying levels of computer experience. *Behaviour & Information Technology, 23*(4), 247–264.

Jacko, J. A., Moloney, K. P., Kongnakorn, T., Barnard, L., Edwards, P. J., Leonard, V. K., Sainfort, F. & Scott, I. U. (2005). Multimodal feedback as a solution to ocular disease-based user performance decrements in the absence of functional visual loss. *International Journal of Human-Computer Interaction, 18*(2), 183–218.

Kaber, D. B., Wright, M. C. & Sheik-Nainar, M. A. (2006). Investigation of multimodal interface features for adaptive automation of a human-robot system. *International Journal of Human-Computer Studies, 64*(6), 527–540.

Kantowitz, B. H. & Knight, J. L. (1976). Testing tapping timesharing: II Auditory secondary task. *Acta Psychologica, 40*(5), 343–362.

Keates, S. & Robinson, P. (1999). Gestures and multimodal input. *Behaviour & Information Technology, 18*(1), 36–44.

Kobus, D. A., Russotti, J., Schlichting, C., Haskell, G., Carpenter, S. & Wojtowicz, J. (1986). Multimodal detection and recognition performance of sonar operators. *Human Factors, 28*(1), 23–29.

Kohlrausch, A. G. & van de Par, S. L. (2005). Audio visual interaction in the context of multi-media applications. In J. Blauert (Ed.), *Communication acoustics.* Berlin: Springer.

Larson, J. (2006). *Common sense suggestions for developing multimodal user interfaces (W3C Working Group Note).* Retrieved December 10, 2006, from http://www.w3.org/TR/2006/NOTE-mmi-suggestions-20060911/.

Liu, Y. C. (2001). Comparative study of the effects of auditory, visual and multimodality displays on driver's performance in advanced traveller information systems. *Ergonomics, 44*, 425–442.

Michon, J. A. (1985). A critical view on driver behavior models: What do we know, what should we do? In L. Evans & R. Schwing (Eds.), *Human behavior and traffic safety* (pp. 485–520). New York: Plenum Press.

Navon, D. & Gopher, D. (1979). On the economy of the human-processing system. *Psychological Review, 86*(3), 214–255.

Nigay, L. & Coutaz, J. (1993). A design space for multimodal systems – concurrent processing and data fusion. In *INTERCHI '93: Proceedings of the Conference on Human Factors and Computing Systems* (pp. 172–178). New York: ACM Press.

Norman, D. A. (1986). Cognitive engineering. In D. A. Norman & S. W. Draper (Eds.), *User centered system design: New perspectives on human-computer interaction* (pp. 31–61). Hillsdale, NJ: Lawrence Erlbaum Associates.

Norman, D. A. & Draper, S. W. (1986). *User centered system design: New perspectives on human-computer interaction.* Mahwah, NJ: Lawrence Erlbaum Associates.

Oakley, I., McGee, M. R., Brewster, S. A. & Gray, P. D. (2000). Putting the feel in 'Look and Feel'. In *Proceedings of the SIGCHI conference on human factors in computing systems* (pp. 415–422). New York: ACM Press.

Oviatt, S. L. (1994). Interface techniques for minimizing disfluent input to spoken language systems. In *Proceedings of the SIGCHI conference on human factors in computing systems: Celebrating interdependence (CHI'94)* (pp. 205–210). New York: ACM Press.

Oviatt, S. L. (1997). Multimodal interactive maps: Designing for human performance. *Human-Computer Interaction, 12,* 93–129.

Oviatt, S. L. (1999). Ten myths of multimodal interaction. *Communications of the ACM, 42*(11), 74–81.

Oviatt, S. L. (2003). Multimodal interfaces. In J. A. Jacko & A. Sears (Eds.), *The human-computer interaction handbook: Fundamentals, evolving technologies and emerging applications* (pp. 286–304). Mahwah, NJ: Lawrence Erlbaum Associates.

Oviatt, S. L., Coulston, R., Tomko, S., Benfang, X., Lunsford, R., Wesson, M. & Carmichael, L. (2003). Toward a theory of organized multimodal integration patterns during human-computer interaction. In *Proceedings of the 5th international conference on multimodal interfaces* (pp. 44–51). New York: ACM Press.

Oviatt, S. L., DeAngeli, A. & Kuhn, K. (1997). Integration and synchronization of input modes during human-computer interaction. In *Proceedings of the SIGCHI conference on human factors in computing systems* (pp. 415–422). New York: ACM Press.

Oviatt, S. L. & Kuhn, K. (1998). Referential features and linguistic indirection in multimodal language. In *Proceedings of the international conference on spoken language processing* (pp. 2339–2342). Sydney: ASSTA, Inc.

Oviatt, S. L., Lunsford, R. & Coulston, R. (2005). Individual differences in multimodal integration patterns: What are they and why do they exist? In *Proceedings of the SIGCHI conference on human factors in computing systems* (pp. 241–249). New York: ACM Press.

Reeves, L. M., Lai, J., Larson, J. A., Oviatt, S. L., Balaji, T. S., Buisine, S., Collings, P., Cohen, P. R., Kraal, B., Martin, J. C., McTear, M., Raman, T. V., Stanney, K. M., Su, H. & Wang, Q. Y. (2004). Guidelines for multimodal user interface design. *Communications of the ACM, 41*(1), 57–59.

Reichenauer, A. (2005). *LUCIA: Development of a comprehensive information architecture process model for websites.* Unpublished dissertation, University of Regensburg, Germany.

Richards, M. & Underwood, K. (1984). Talking to machines: How are people naturally inclined to speak? In E. D. Megaw (Ed.), *Contemporary Ergonomics 1984* (pp. 62–67). London: Taylor & Francis.

Rosson, M. B. & Carroll, J. M. (2001). *Usability engineering: Scenario-based development of human-computer interaction.* San Francisco: Morgan Kaufmann.

Rosson, M. B. & Carroll, J. M. (2003). Scenario-based design. In J. A. Jacko & A. Sears (Eds.), *The human-computer interaction handbook: Fundamentals, evolving technologies and emerging applications* (pp. 1032–1050). Mahwah, NJ: Lawrence Erlbaum Associates.

Schomaker, L., Nijtmans, J., Camurri, A., Lavagetto, F., Morasso, P., Benoît, C., Guiard-Marigny, T., Le Goff, B., Robert-Ribes, J., Adjoudani, A., Defée, I., Münch, S., Hartung, K. & Blauert, J. (1995). *A taxonomy of multimodal interaction in the human information processing system. Multimodal integration for advanced multimedia interfaces* (Report of the Esprit Project 8579 MIAMI No. WP 1). Nijmegen, Netherlands: University of Nijmegen.

Seagull, F. J., Wickens, C. D. & Loeb, R. G. (2001). When is less more? Attention and workload in auditory, visual and redundant patient-monitoring conditions. In *Proceedings of the Human Factors Society 45th Annual Meeting* (pp. 1395–1399). Santa Monica, CA: Human Factors and Ergonomics Society.

Spence, C. & Driver, J. (1997). Cross-modal links in attention between audition, vision, and touch: Implications for interface design. *International Journal of Cognitive Ergonomics, 1*(4), 351–373.

Spence, C. & Driver, J. (1999). Multiple resources and multimodal interface design. In D. Harris (Ed.), *Engineering psychology and cognitive ergonomics: Transportation systems, medical ergonomics and training* (Vol. 3, pp. 305–312). Aldershot: Ashgate.

Todd, J. W. (1912). *Reactions to multiple stimuli* (Archives of psychology, No. 25). New York: Science Press.

Tzelgov, J., Srebro, R., Henik, A. & Kushelevsky, A. (1987). Radiation search and detection by ear and by eye. *Human Factors, 29*(1), 87–95.

Van Laer, J., Galanter, E. H. & Klein, S. J. (1960). Factors relevant to the development of aircraft warning and caution signal systems. *Aerospace Medicine, 31*, 31–39.

Vilimek, R. & Hempel, T. (2005a). Effects of speech and non-speech sounds on short-term memory and possible implications for in-vehicle use. In *Proceedings of the 11th International Conference on Auditory Display ICAD 2005* (pp. 344–350). Limerick, Ireland: ICAD.

Vilimek, R. & Hempel, T. (2005b). Eine nutzerzentrierte Analyse von Erfolgsfaktoren zur Sprachbedienung im Automobil. *Forum Ware, 32*(1–4), 47–51.

Vilimek, R., Hempel, T. & Otto, B. (2007). Multimodal interfaces for in-vehicle applications. In J. A. Jacko (Ed.), *Human-Computer Interaction, Part III, HCII 2007, LNCS 4552* (pp. 216–224). Berlin: Springer.

Vilimek, R. & Zimmer, A. (2007). Development and evaluation of a multimodal touchpad for advanced in-vehicle systems. In D. Harris (Ed.), *Engineering psychology and cognitive ergonomics, HCII 2007, LNAI 4562* (pp. 842–851). Berlin: Springer.

Vitense, H. S., Jacko, J. A. & Emery, V. K. (2003). Multimodal feedback: an assessment of performance and mental workload. *Ergonomics, 46*, 68–87.

Vogels, I. M. L. C. (2004). Detection of temporal delays in visual-haptic interfaces. *Human Factors, 46*(1), 118–134.

Wickens, C. D. (1984). Processing resources in attention. In R. Parasuraman & D. R. Davies (Eds.), *Varieties of attention* (pp. 63–102). London: Academic Press.

Wickens, C. D. (2002). Multiple resources and performance prediction. *Theoretical Issues in Ergonomics Science, 3*(2), 159–177.

Wickens, C. D. & Hollands, J. G. (2000). *Engineering psychology and human performance* (3rd ed.). Upper Saddle River, NJ: Prentice Hall.

Zoltan-Ford, E. (1991). How to get people to say and type what computers can understand. *International Journal of Man-Machine Studies, 34*(4), 527–547.

Chapter 7

Did you find yourself hesitating for a second here, wondering what branding issues might have in common with the other chapters in the book?

Well, this is where it gets real: voice dialog systems live in the real world, so they not only need to be easy to use but are typically set up by a company for a certain purpose. As a natural consequence, this new instrument has to reflect the overall identity of the company. It's as easy as that: a commercial voice dialog system is per se a communication tool and must therefore follow the rules of branding and corporate identity.

Keeping this in mind, leave the rest to Carl-Frank.

Prof. Carl-Frank Westermann

Professor Carl-Frank Westermann joined MetaDesign in September 1996. He has been in charge of developing and managing its sound branding unit since 2001, working with clients such as Allianz, Lufthansa, Siemens, eBay, Audi and Volkswagen. Westermann has a degree in business administration and studied, among other things, organizational psychology and music. Since 2002 he has been a guest professor at the Berlin University of the Arts, where he is responsible for acoustic conceptualization in "Sound Studies", an advanced studies program.

Carl-Frank Westermann believes in the holistic presentation of brands and multi-sensory brand communication. "Today companies are usually aware of how important visual design is for their brand image. Yet corporate sound is rarely a part of brand strategy – even though sound has a direct impact on human emotion. Corporate sound sharpens brand image and enhances recognition."

Carl-Frank Westermann is Creative Director of Sound Branding at Meta-Design AG.

7 Sound Branding and Corporate Voice – Strategic Brand Management Using Sound

Carl-Frank Westermann
MetaDesign AG, Berlin, Germany

Developing the value of a brand requires a long-term communication strategy that builds and hones a unique brand personality.

Whereas visual forms of expression have captured a great deal of attention in this context, sound was neglected for a long time. Only recently have sounds, music and voice attracted greater interest. Of all our senses, hearing is the one that we are least conscious of in our daily lives, and it is for this reason that the auditory dimension of strategic brand management has previously been overlooked. There is huge difference between our conscious perception of sound and its true significance. But what special impact can sound branding have, and what benefits does the auditory realm have for brand identity?

7.1 The Importance of the Auditory Dimension

It is beyond dispute that hearing is a constant feature of our lives, one that provides guidance, support and balance every second of the day. The findings of perceptual psychology even show that, when it comes to delivering messages, sound has an advantage over visuals. Our sense of hearing is based on sound waves penetrating directly to the brainstem. It absorbs information more quickly and allows us to find our bearings at an unconscious level. We cannot simply "turn it off" in the same way we can avert our glance. We cannot stop hearing.

Professionally produced audio brand identities can already be heard in individual applications of a temporary nature such as communication campaigns, promotions and some products. But at the time of this writing, the implementation of audio brands often lacks continuity and consistency when it moves beyond the boundaries of individual applications. This is especially true of the voices that are used. Even so, the same laws apply to sound branding as to consistent visual design.

7.2   The Sound of a Brand

When developing the sound of a brand – the corporate sound – it is crucial not to choose a sound that is pleasant or appealing primarily for faddish reasons. A well-conceived audio identity must be based on the sonic equivalents of brand values. Proceeding from these sonic equivalents, sound branding aims to develop sound creations that are derived from the details of brand values and that characterize the brand. The final audio results used to define a brand must contain a story about the brand's origins that is as credible as possible. This ensures a comprehensible uniqueness and an enduring acoustic foundation (brand-sound connotation) for corporate sound as part of the interplay between all the multi-sensory factors of integrated brand communication.

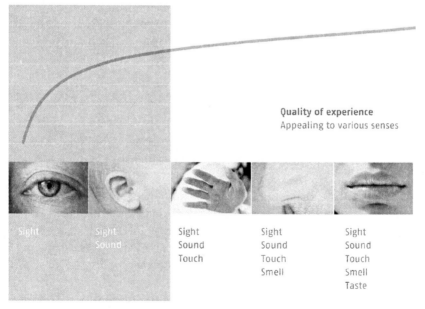

Source: 5-Sense Branding Principle, different/MetaDesign 2005

Fig. 7.1. Brand and Multisensory Perception

7.3   The Sonic Equivalents of Brand Values

A sonic brand requires an underlying story that is so compelling that people enjoy and take pride in telling it. It must also be a story that others understand, enjoy hearing and respond to with enthusiasm. To find this story, the sound branding process begins by examining and selecting the sonic equivalents of brand values. Once these have been defined, brand-specific influential factors such as history, guiding ideas, service/product details, cultural links, etc., can be harnessed to transform the sound of a brand into a specific corporate sound. All the methodological steps necessary to develop corporate sound are rooted in an audio concept that, in turn, provides a foundation for all briefs on concrete audio applications.

7.4   The Building Blocks of Corporate Sound

Corporate music, corporate soundscape and corporate voice are the strategically based building blocks of corporate sound. These elements can be used to design a great variety of applications based on individual requirements. Depending on the application, the elements of corporate sound can be weighted differently. If companies succeed in using corporate sound consistently in an ongoing process across all the brand's audio touchpoints,

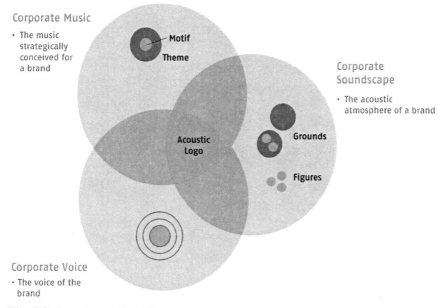

Fig. 7.2. Sound Branding Elements
Source: Sound Branding Elements/MetaDesign 2006

they will establish a high degree of brand recognition. The sound branding process provides the brand with a communicational, emotional tool in the world of multi-sensory perception. One element of corporate sound whose effect is often underestimated is the human voice.

7.5   The Power of the Voice

The sound of a voice influences communication in a very special way. A product of evolution and socialization, the sound of a voice is precisely analyzed by the listener for its information and communication content. It is nearly impossible to suppress this decoding process – as becomes evident when cultural practices diverge from the familiar and cannot be deciphered immediately, or when the language spoken is not understood.

Analyzing and classifying the information carried by the voice is one of the most important skills in human interaction. The sound of a voice triggers a series of complex associations in the listener, allowing him or her to "recognize" the speaker or to form an idea about that person. These comprehensive processes are mostly unconscious. An astonishing fact is that human beings have the capacity to form complex mental images based on very little information, particularly when they lack the visual information corresponding to aural phenomena. A comparable process is our purely visual perception of a person's stature, which also expresses certain personality traits. To a certain extent, our perception of these traits may vary, yet stature remains a characteristic perceptual feature of that person.

7.6   Brand-Specific Application

Corporate voice is currently attracting increased attention, yet it is still treated as a marginal topic. Brand managers who wish to learn more about it often question the economic sense of this brand management tool. Without wanting to suggest this questioning is unjustified, I believe it is worthwhile to analyze why brand managers no longer question the efficacy of visual forms of corporate design. After all, there are currently no studies of either audio or visual brand communications that can validly prove that a euro spent on corporate design or corporate sound has a positive effect after implementation (never mind measuring that effect). Even the most reliable market research methodologies are incapable of precisely answering this question, which touches upon highly complex impact factors of brand perception. That said, key financial figures lead to the clear conclusion that multi-sensory communication measures taken in advance to increase brand value do indeed pay off (including a focused strategy, corporate communications, design, sound). Since brand development processes are long term in nature, brand managers should from the outset rely on their own views and understanding and on the trust they place in the value of brand-building investments.

The human voice has an impact that goes far beyond the meaning of words. Not only does the quality of a voice influence the listener's level of attention and willingness to take in information. It also has a strong emotional impact. Thirty-eight percent of the effect we have on other people can be attributed to our voice, only 7 percent to the words we utter, and a whopping 55 percent to our body language. As far as brand personality goes, we can see the enormous potential of designing this voice, the corporate voice, to reflect the brand.

A conceptually developed voice can make an immense contribution to sonically defining brand identity. A concordant relationship between brand personality and the voice's expressive character is a decisive factor for consistently and credibly delivering a brand message.

As a "product of nature", the human voice eludes strict definition, yet concepts from music theory can to a certain extent be applied to the sound of a voice. This means that the voice and its use as an element of corporate sound can be controlled to reach specific objectives.

An integrated approach to corporate sound not only guarantees a brand-specific implementation of corporate voice. It also ensures that a coherent overarching perception grows out of subsequent combinations of this voice with a corporate soundscape and corporate music.

Used in brand communications, the voice transfers the effects of its "personality" to the brand. However, the effects of a voice are not equally powerful in all media in the different branches of brand communication. Focused, brand-specific usage requires companies to take into account the goal of the application and the context of the target group. In purely auditory channels such as the telephone or radio, a voice has much greater force than in audio-visual presentations. In such cases, sensory modalities are tied to the audio channel. Furthermore, there is no further information to assist interpretation.

Corporate sound takes brand personality into sonic dimension. Additional applications that can be developed from corporate sound are based on a holistic concept of brand identity. The corporate sound brief defines the framework for the creative work and clearly formulates the objectives. What is at stake here is not a rigid structure of design rules, but a conceptual foundation that can be flexibly transformed into the respective application – and that, above all, contributes to efficient work. An application designed in this way conveys brand identity through the audio channel, making it recognizable and differentiating the company from competitors.

7.7   Vocal Character

The telephone is a purely auditory channel that is currently experiencing a renaissance thanks to sophisticated new applications. The large number of contacts facilitated by the phone and their relatively long duration make it a medium that should not be neglected in terms of audio design. Technological developments have given rise to interactive voice systems that are increasingly supplementing and replacing call centers. This means that a single voice is being used to influence brand perception. Efforts to design the character of this voice, the so-called persona, are assuming greater importance for the success and acceptance of such applications. An additional point is that customers are addressed directly.

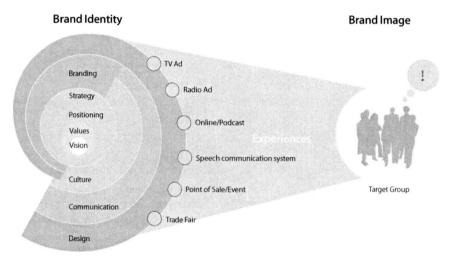

Fig. 7.3. Voice Touchpoints
Source: Brand Identity/MetaDesign 2006

7.8   Voice Applications

Verbal interaction in electronic systems represents a huge technological challenge, and technological developments have taken a quantum leap over the last decade. Even so, voice applications suffer from image problems. Now that technologies have matured, the next goal must be to create distinguishing features. Besides technical and dialogue design, one central characteristic of voice dialog systems is the human voice. All of these factors taken together determine the user's impression and customer acceptance. The "staging" of the human voice is also taking on growing importance in this field. Manufacturers and providers no longer set store by technical expertise alone, but also by identity.

Companies with a corporate sound program of their own possess the necessary tool to design voice applications or to have others do the work. The process begins with guidelines for the selection of male or female speakers and covers presentational criteria for the desired expressive character and sound, as well as musical elements that provide a framework for the system. A conceptually developed corporate voice and its consistent, integrated use as a basic element of corporate sound can provide orientation for both providers and users of voice applications. Most importantly, it can intensify the feeling associated with a brand.

An examination of present-day brand communications shows that there are hardly any applications that get by without a voice. From commercials to voice-user interfaces, brands speak to and with customers. The voice incorporated into such applications always has an effect. Corporate sound contributes to making this effect consistent with brand personality. Grasping corporate voice as an instrument of brand communication – one that conveys both information and brand feeling – will herald the next stage in the evolution of voice applications as tools of communication.

References

Jochen Bonz (Hrsg.), Sound Signatures. Frankfurt am Main 2001.

Kai Bronner, Rainer Hirt (Hrsg.), Audio-Branding: Entwicklung, Anwendung, Wirkung akustischer Identitäten in Werbung, Medien und Gesellschaft. München 2006.

Michael H. Cohen, Voice user interface design. Boston 2004.

V. Fischer, A. Hamilton, Theorie der Gestaltung, Frankfurt am Main 1999.

Daniel M. Jackson, Sonic branding: an introduction. London 2003.

Mark Lehmann, Voice Branding: Die Stimme in der Markenkommunikation. Berlin 2007.

J.F. LePla, L.M. Parker, Integrated Branding: Becoming Brand-Driven Through Companywide Action. London 1999.

Martin Lindstrom, Brand Sense: how to build powerful brands through touch, taste, smell, sight and sound. New York 2005.

Robert Remez (Author), Handbook of Speech Perception. Oxford 2005.

K.R. Scherer (Hrsg.), Vokale Kommunikation: Nonverbale Aspekte des Sprachverhaltens. Weinheim 1982.

Manfred Spitzer, Musik im Kopf: Hören, Musizieren, Verstehen und Erleben im neuronalen Netzwerk. Stuttgart 2007.

Chapter 8

Finally, Paul advises the reader not to overlook the probably most neglected 'target audience' when designing speech dialog systems: our deciders within the own organization. With bright anecdotes and witty footnotes from his professional life as a decade-long speech user interfaces developer, Paul addresses some issues that are known to many of us but taking a view from the far side – always with a wink in his eye.

Paul Heisterkamp, M.A.

Paul Heisterkamp received an MA in German philology, philosophy and general linguistics at Münster University, Germany. Starting out in 1987 with the then AEG research at Ulm, Germany, that later became Daimler corporate research, he worked on numerous national and international research projects on spoken dialogue systems, such as, e. g. Esprit-I P26, SUNDIAL, DISC, or DARPA Communicator, acquiring expertise in all aspects of these systems, as well as on company-internal projects. His current focus is on multimodality and system integration in mobile environments. He authored and co-authored more than 30 papers presented at international conferences and has been an invited lecturer at several universities in Europe and the US.

8 Speak to the Target Audience – Some Observations and Notes on the Pragmatics of Spoken Dialog Systems

Paul Heisterkamp
Daimler AG, Ulm, Germany

8.1 Introduction

In this book, most chapters deal in one way or another, with techniques to build and improve spoken dialogue systems that deliver the best possible service to their users. The sub-title of this collection implicitly assumes that the people who call in to telephone-based services or who use multi-modal interaction with mobile systems *are* this target audience. And, of course, under scientific, technical and engineering aspects, this is undoubtedly true.

In real life, however, this assumption is not always correct and some-times even misleading. Rather, under economical, political and practical aspects, the first target audience, and, in some cases the most decisive one, are the people in those organizations who deploy these systems, i. e. make them available for their customers. These people, first of all have to buy the dialogue system and its design. They pay (out of their budget) for its development, they allow it to be integrated in their company's workflow, or offer it as a product or service to their company's customer.

Of course, these people want a good system that helps them with their business goals, but: they also want to have a word in it – and sometimes quite literally: This is what makes dialogue and interaction design special in some respects. While few decision makers have actual experience with many technical details of complex systems, and rarely have a personal preference as to whether you use one or the other hardware or programming language, they all can speak and do conduct dialogues and do have and do pronounce personal preferences.

This chapter is not a scientific contribution. To the best of our knowledge, no scientific examination of the practicalities of deploying spoken dialogue systems exists. This contribution is based on the author's experience and the exchange with colleagues over a substantial period of time. This chapter is not aimed at having us dialog designers and engineers indulge in mutual 'dilbertism', slightly self-pitying anecdotes of what went wrong because of 'them' – managers, accountants and all those influential people who, by virtue of being able to speak, have an 'opinion' of what a good Voice User Interface should be like – (although more exchange in the community on these issues not only can not be avoided, but even would be a desirable side-effect).

Rather, we try here to raise the awareness of Spoken Dialogue System designers early on to the fact that about every technical system that makes it to the market is, in many ways, a compromise. Time, cost, supplier deadlines, bad weather – you name it – influences of all of these kinds are constantly driving the product away from the originally intended form, and it is part of the everyday life of engineers and product managers to twist and turn their plan in order to come as close to the original design as possible. This task is made easier if all the parties involved agree on a common goal: to create, build, sell … the best possible compromise considering all the influences. Therefore, it is good to know, and to invest some time and effort in finding out, who the parties involved really are: Who is my first 'target audience'?

One of the tasks of dialogue designers is to convince this audience of those decisions in the dialogue design that do not conform to this audience's initial preferences. This means, first and foremost, that we also have to listen to this audience: What lies at the base of their wishes? And: How can I prepare and formulate arguments for the 'better' design such that it is accepted? This chapter addresses these questions.

Bear in mind, however, that we address these questions in the context of commercial dialogue systems. While conversational dialogue might be a desirable goal, and a very interesting research topic, "…naturalness and freedom of expression are not necessarily a prerequisite for usability, given the constraints of the current technology" (Pieraccini & Huerta, 2005).

8.2   A Long, Long Time ago in a far Away Galaxy…

…(in fact, this was in the early nineties…) a car insurance company asked a team of researchers and developers (…when there was virtually no speech industry around) to build a spoken dialogue system that would offer price quotes to people who called in. The task was described as 'to elicit all relevant personal information from the caller so that the (insurance company's) calculation system can immediately provide a reliable quote for both liability and full body insurances'. The project was started by a tech-wise assistant call centre manager who had convinced his immediate superior that speech and dialogue technology had advance far enough to grant the specialized direct insurance company (a low/cost subsidiary of a major insurer) more competitive edge.

As a description of what the spoken dialogue system should do, the speech team received an application form sheet, the one that people received who wrote to the company for information. The potential applicants were asked to fill in information like their car's make, model, motorization, their birth date, marital status and so on. The dialogue designers happily went to build an initial prototype system.

Some time into building that system, which for the technology at that time was quite an ambitious enterprise, the dialogue people found out that apparently there was some inherent logic in the questionnaire, as some questions were dependent on the answers to other questions – and thought that this could be useful to optimize the dialogue by reflecting this logic and not ask everybody everything. The insurer's answer to this was simple: "No, you can't do that. We won't disclose our business logic and our calculating principles neither to you nor to our callers. The way we calculate our rates is our major competitive instrument. We regularly plant fake applications with our competitors, as they do with us, to find out how they calculate their prices – no way you can reduce the number of questions asked to the caller such that our competitors could infer some of our calculating principles."

Next, the dialogue development team wanted to gather some live data on the calls handled in the company's call centre. Mind, this was in the days when insurances still owned and operated their call centers themselves and not everything was outsourced yet. The call centre manager agreed and told his employees in a quite direct fashion that from now on their calls would be recorded "to prepare for automatization". This was the first time the call centre operators officially heard of the project. Consequently, the trade union representative stormed into office of the call centre manager's boss and yelled at him about job losses and lay-offs and so on. The higher-ranking manager was completely shocked by this, as he assumed, rightly, that the speech dialogue system was planned specifically to handle only call overflow as it regularly happened in the three minutes after a TV advertisement. In any case, to avoid quarrel with the trade union and to calm down the employees, the manager decided on the spot that no recordings would take place. Later, however, he permitted the dialogue designers to talk to call centre operators on a volunteer basis. Regrettably (or understandably), few of the still quite aroused operators volunteered and fewer still had their calls recorded – far too few to create anything like a database, and only giving anecdotal evidence of what really happened in the calls – just enough to show that successful dialogues more often than not deviated considerably from the dialogue from sheet the operators should nominally follow.

In the course of the discussions of these events, the dialogue designers were made aware for the first time of the 'overflow' situation. The problem the insurers really had was the following: Immediately after a TV broadcast of their advertisement, people would dial the number displayed on the screen, so many, in fact, that not even two additional call centers could possible handle but about half of these calls. As market studies had revealed, within three minutes the callers who did not get through would hit the re-dial button no their phones, If again they were put on hold or received a busy signal, they would not call that particular number again for the next three months – invalidating the company's spending for the expensive airing of the advertisement.

So, from the point of view of the marketing and sales people at the insurance, the task of the spoken dialogue system was first and foremost, to keep callers in the line and keep them happy and occupied until a call centre agent became available again, i. e. for an average time of 3 minutes 16 seconds. They wouldn't *mind* if the spoken dialogue also collected some useful data that would help the agent speed up the quote-providing human-human dialogue, but they didn't really care either.

Before the dialogue people could set out to adapt their system to these new requirements, however, the story ends: at about that point in time, the car insurances mother company was taken over by a competitor, the business was transferred to that competitors respective subsidiary, and the dialogue project found another user and eventually became quite successful. So, there are no stories here about finding a voice talent, quarrelling over prompt wording or emphasis, budget cuts or other things that make life interesting.

8.3   Identify Key Players

Who are the people in your customer's organization that are really important? Directly, the designers and developers are dealing with the project managers and technical people at the deployers' side. Now, quite often, the people who are in immediate contact with the technology, and whom technology helps to achieve their goals, are on the side of the developers and designers. They will rarely press through changes that endanger the performance of the system. However, just as rarely, the project managers and techies have the final word. They are bound into their company's hierarchy.

In the fairy tale above, the speech and dialogue team was quite naïve. They believed that their immediate points of contact on the insurer's side were actually the people who decided it all. In fact, they did decide much – but not everything. They had above them and around them many others who also had some stake in the game. Now, some of the troubles that particular project did run into were due to the tech people within the insurance company not fully realizing who was or might be concerned (and yes, they were not too skilful in handling the internal communication as well). Today, a hopefully less naïve dialogue team would see it as part of analyzing the task to ask questions beforehand: "How will this spoken dialogue system integrate with the workflow of your organization?" "Which of your internal processes are involved and who are the owners of those processes?" "How are your customer relations people involved in the speech project and do we need to pay special attention to – say – branding and corporate identity issues?" etc.

With good answers to questions such as this, one can ask the deployer's tech people to invite the key players to meetings and demos early on. Mind you, this does not assure you'll not get into trouble with some of them, but having a higher-ranking decider supervise the decisions in such a project reduces the risk of getting caught between the fronts of intra-corporate trench wars. And you want decisions. It is the managers' task to make decisions, reluctant as some may seem to do so. You want decisions early on in the project that hopefully will be stable and also can help you to fence off attempts to influence the project from the sidelines.

While the above holds true for just about every industrial project, in spoken dialogue systems you have to live with the fact that even non-technical decision makers have an opinion on the subject matter.

Sometimes, even, the most important people are very hard to find. A long time ago, a company had a prototype system running where people could speak digit strings (order numbers). The project team proudly and successfully demo'ed the prototype to the CEO and gave him the internal phone number. The CEO complained after a few days that the recognition rate on the digit strings was not sufficient. The project lead tried to make the best of it and asked: "Oh, did you try it out?" "No, but my wife did, and it doesn't work for her – you'd better do something about that, I won't accept it the way it is…" Inspection of the log files (good there were any) revealed that the CEO's wife apparently knew the company's order numbers by heart and did not listen to the prompt asking for the numbers in groups, as in written form they were separated by blanks into groups of three digits. Rather, with her very high-pitched voice she rattled of the whole 12-digit string in one breath, the energy level consequently dropping towards the end and often under the segmentation threshold. As it was out of the question to tell the CEO's wife to behave and listen to the prompts, eventually the project team decided to lower the segmentation threshold considerably for the prototype – and consequently, the lady succeeded in getting her order numbers through no problem. Though the recognition rate dropped for everybody else, the CEO came back and said: "Well done, now the system may go live!", upon which the segmentation was reset to the original level and the system went into production, performed beyond all expectations, and, if it has not been replaced, still serves the company well to this very day…

The point here is that the CEO, the decider, had, in some weird way, his own evaluation criteria. He considered his wife to be a representative sample of his 'target audience'. What saved the project was the dialogue team being able to track down these criteria and adjust their system such that it could meet them. If the CEO hadn't disclosed whose impression of the system it was that he based his decision upon, the project would have been doomed.

8.4   Understand Decision Makers

What do the decision makers want and what lies at the base of their wishes? Managers and project leads on the deployer side want to decide, want to make a contribution, want to have at least al little bit of their own represented in the system that they can point to and say: "I did this!" – and this the more as they do assume that their understanding of what dialogue is, what prompts should be like etc. is as good as everybody else's. Moreover, it is in general their job and their task to decide, and, rather often, they are forced to decide on something they have little personal knowledge of. Small wonder that when it comes to something they think they understand, like spoken dialogue, they want to make decisions and bring in something of their own.

The point is not that managers aren't allowed to make contributions, make decisions and even make bad decisions. Provided they have somehow obtained the relevant decision criteria and take the responsibility for what they decide, this is fine. If they are aware of the consequences of their decisions in terms of success of the overall system, this is even better. In our field, however, more often than not their contributions and decisions are based on what can best be described as personal taste and some acquired ideas and/or prejudices.[1] They come up with some weird ideas:

- e. g. insist on speech as 'automatic crutches' for IVR systems: "For Sales say 'One', for Support say 'Two', or say 'Three' if you want to be put in the waiting queue for an operator.", instead of: "Do you want Sales, Support or an Operator?";
- e. g. insist on the system using technical slang in prompts rather than the common language expressions people are much more likely to understand, like 'LED' instead of 'little red light';
- e. g. insist on some kind of 'over-politeness', littering the prompts with 'please'es, 'could you be so kind as to's and all that stuff that gets on users' nerves really soon;
- e. g. insist on the voice talent for the prompt recording to either use a completely exaggerated or – conversely – an extremely dull tone of voice, in any case not the appropriate one for the task...

While there are managers (the sceptics) who outright say: "Personally, I hate the idea of talking to a computer" or: "It didn't work when I last saw it fifteen years ago, so why should it work now?" and are very reluctant all through the project that has somehow fallen into their hands as to what the speech dialogue system is capable of and what it should do, others (the believers) fall into the opposite extreme. They can't believe that speech systems have any limitations at all. There are those who get along very well with their dictation system, and those who've seen it all in '2001 – A Space Odyssey' (or worse, on 'Knight Rider'). Both tend to want to overload spoken dialogue systems with interpretative capacity and functionality requirements current technology cannot possibly meet. And both tend to be quite disappointed when speech engineers point to the difference between dictation geared towards visual control and post-processing of speech recognition results and constructing meaning for meaningful actions out of a speech signal, or simply the difference between realistic vocabulary sizes and science fiction.

The problem here is that it is hard to convince these decision makers that their opinions and tastes will result in sub-optimal system performance without presenting a full-blown study to prove it. And even then (cf.

---

[1] This the case for other human-computer interfaces as well: Allegedly, it was the personal decision of the chairman of (another) major automotive manufacturer to adopt a red-blue color scheme for the instrument cluster of a whole generation of their vehicles. The *proven* fact that this design made the display very hard to read for color blind people (about 3% of the population, give or take a few) did not deter the company from even widely advertising their cool 'underwater design' – at least they stuck to that decision!

footnote 1), this calls into question the personal opinion of the decision maker as well as his or her decision competence on the subject – both of which you want to avoid, as you loose either way. Remember, this is the first target audience, and it has to be listened to. So, what can we do not to fall into this trap?

8.5   Encourage Decisions

As the 'speech industry' matures, the developers and designers rarely have initial direct contact with the deployers. It is the task of the sales people on the side of the speech enterprises to find potential customers and get them interested in their product or service. And it is the task of the sales people to offer to the customers what the customers want (or think they want) for the lowest (or highest, depending on which way you look) possible price. Quite often, the developers get stuck between a rock and a hard place in trying to reconcile the demands as negotiated between their sales and the customer with the budget they get allocated.

So far, so bad – that's how it goes everywhere. In speech industry, you get the deployers managers' personal opinion on top of that! What can you do about it?

Manage expectations: The first task is to include the sales people of your company as closely as possible (sure, they always are away on cus- tomer appointments) into the internal communication loop. This is not the case in each and every company. Not only should sales be aware of what the techies can do (or can not do), development should also try to find out as much as possible of what may be in the pipeline for them in terms of customer demands. The goal here is to try and prevent sales from 'over- selling' spoken dialogue systems to the 'believers', whilst providing with convincing example installations and fresh and stable demos for the scep- tics. If we make a strong point as to what we can or can not do for a given budget, then all the extras can be tagged with a price.[2]

Be cost-aware: While decision makers may be quite touchy as far as criticizing their opinions and beliefs are, they also (mostly) are very aware of the financial aspects of their decisions. In many companies the devel- opment budget is not directly related to the savings or revenue side, i. e. the person out of whose budget the spoken dialogue system is financed does

---

[2]   Some companies (e.g. SpeechCycle http://speechcycle.com/why/improve.asp) already base their business in continuously adapting their systems and improv- ing caller success rate and caller satisfaction, hardly doing any customer tuning at all. They get paid per successfully handled call, so they don't sell a spoken dialogue system, but offer a service.

not necessarily benefit from the added value it (ideally) generates. That is why for some decision makers, the system's overall performance is secondary to them having their mark stamped upon it. Consequently, in most cases the best arguments one can have to prevent decisions that would deteriorate system performance are arguments related to cost rather than escaped revenue. If your specifications and the expectations of your sales people permit it, for every decision or proposal you don't agree to, you can raise a sign with a money amount on it and see if the proposal is maintained.

Offer alternatives: Managers want to decide (cf. above). So, you give them something to decide over. And if that is just a yes/no, take-it-or-leave-it decision, you'll get something like "Naw, I don't like either..." – or, worse, a fit of creativity that might spoil the coherence of your design, flaw the system and endanger its success. So, it is better to prepare at least two options that fit in the system design. Make sure you get a chance to present them and have the pro's and con's ready to offer as a basis for the manager's decision.

A slight more deviant version of the process to offer alternatives is to deliberately build more or less obvious flaws into the system. You can start with simple ones like typing errors in system prompts, inducing garbage words in some prompts. Make sure your demo includes these flaws, and also make sure that none of you nor your team lets out the least sign that they noticed the flaw. Wait for the manager to notice it and point it out to you. Then, rather than falling into excuses, praise the manager for his/her sharp attention and say that without his or her help this would possibly have gone into production and you can't express your thanks enough that the manager prevented you from putting such a flawed system out into the world. It may seem quite a cheap trick, but it has worked in surprisingly many cases, and continues to do so...

Be affirmative: If you get confronted with a fit of managerial creativity, most of the time it is not a good idea to argue against it. You don't want to try to talk the decider into admitting, possibly in front of their own entourage, that the idea they just proposed might, all things considered, not be so good after all. It is usually a more promising approach to accept the notion of the proposal and say something positive but non-committing like "This is an interesting proposal. We will try to build this into the system, but this requires careful consideration and assessment of the mutual influence it has on the other parts of the system..." If you really hate the idea, then, for the next meeting, prepare an argumentation that goes along the lines like 'locally, idea is good; in fact, we consider integrating it in our next generation; globally, introducing it at this point of the project entails that this and this doesn't work any more (or gets a lot more expensive or delayed);

do you want to bear the consequences?' Caution: do not introduce the topic on your own initiative: Managers have little time, and some have a rather short attention span – in more than one case, the bad idea had been completely forgotten by its proposer, and only was introduced into the system because the designers (or their sales people) forgot to forget it – so don't say no, but don't rush either!

Be open: Listen closely to what the decision makers say. In the fairy tale above, if the dialogue designers had had a hunch earlier on that they might have been allowed to achieve the goal of keeping people on the phone for just over three minutes on the average *without* necessarily going through all the grounding etc. of task-oriented dialogue, they would have come up with a whole set of creative ideas as to what an interactive *social* dialogue could be like. Listening to decision makers can, at times, reveal much of the background and the basic intentions of the deployer. Rather than having a narrow view of delivering the best technical solution, this may lead you to find that 'dialogue success' may, under certain circumstances, be measured differently than by common metrics.

8.6   Live with it – but Keep up the Effort!

Regrettably, in some cases none of the above mentioned techniques yield the desired result. We see bad spoken dialogue systems all around us, and, when you mention to laypeople what you're doing you're more than liable that one of the people around you will complain to you about how bad speech systems are – and, mind you, in all of the above, we speak of professional spoken dialogue designers being somehow forced to deliver flawed systems: there are all these systems around where people who have no knowledge in speech and dialogue at all just take an off-the-shelf recognizer and produce yet another complete disaster.[3] Bad systems out in the marketplace endanger the reputation of everyone in the business, endanger the continuous development of the technology and, eventually, our jobs. So, it is not just the designer's pride that is hurt (although every engineer and designer feels a little personal pain if he or she are not to allowed to

---

[3] The author intended to provide a reference here, but rather than risking a law suit asks the reader: Haven't you yet seen systems where people foreign to the field 'just added voice capability', to an already bad user interface for some spreadsheet data entry, only to declare in a paper submitted to one or the other major international HMI conference that 'voice technology obviously isn't mature enough to really be useful'? (Of course, such papers are rejected at decent conferences – but who knows where else they appear?) cf. also Skantze (2007).

deliver the best they can), but it is in the interest of the community that we continue to strive to build systems that work really well.

The fact is that still more and more systems are being deployed, that more people accept and even begin to like these systems and that, overall, even the new bad systems we see are at least a little bit better than last years'; the fact that enterprises continue to introduce successful systems,[4] and that with each one of them, the technology gains ground. So, it is fair to say:

> "I have to admit it's getting better,
> a little better all the time..."[5]

Still, other than seeing to it that every system we turn out into the world is a good as it possibly can be considering the circumstances, there is more we can do. Manage expectations big way is one possibility: a combined speech industry effort in educating customers. This might be a good idea in principle, but as every participant from the speech side would have to point to and lay open the particular limitations of their systems, and as, united tough they may be in suffering from bad systems, they remain competitors and could never be sure the others in speech industry wouldn't try to exploit the admittance of these limitations.[6]

We, the designers, can use a different path. Why don't we all agree to build into every system in a clandestine, covert operation at least the 'Speech Graffitti' functionality (Rosenfeld 2000ff., Heisterkamp 2003): being able to say with a reasonable list of synonyms, at any time in the dialog, something like keyword:value and make sure this works – neither your bosses nor the deployers need to know about this: yes, we don't even

---

[4] Everybody knows at least one; e. g. the by now well established 'Julie' system for Amtrak receives wide acclaim:
http://www.nytimes.com/2004/11/24/nyregion/24voice.html?_r=1&adxnnl=1& pagewanted=1&adxnnlx=1197828065-syCmmPCsvcBkK1shZp9F1w&oref= slogin –
or refer to the winners of the German Voice Awards (http://www.voiceaward.de/). However, the author tends to disagree with the Voice Awards people saying on that page: "...in 2007 there was no more direct developer's contest, as the development on the market, and the developer's competence available in the market by now, only produces good solutions." (translation PH).

[5] Lennon/McCartney. The author apologizes to any company that might potentially have used this song in their advertising for revealing the true continuation. Wikipedia says: 'In response to McCartney's line, "It's getting better all the time," Lennon [cynically, ph] replies, "It can't get no worse!"' (http://en.wikipedia.org/wiki/Getting_Better).

[6] cf. the complaint about 'companies' naïve approach to spoken dialogue projects' (Voice Days 2007).

have to know this between each other! Try out the next foreign system you stumble upon, and see if this approach works there once you got stuck – if it does, then you know we're winning!

Economy (and time) is on our side…[7]

References

Heisterkamp, P. (2003). "Do not attempt to light with match!": Some thoughts on progress and research goals in Spoken Dialog Systems. In *Proceedings of Eurospeech '03* (pp. 2897–2900). Geneva.

Pieraccini, R. & Huerta, J. (2005). Where do we go from here? Research and commercial spoken dialogue systems. In *Proceedings of the 6th SIGdial Workshop on Discourse and Dialogue*. Lisbon.

Rosenfeld, R. (2000 ff.). USI or 'Speech Graffitti' (-tt- sic!), e. g.
http://www.2.cs.cmu.edu/~usi/

Skantze, G. (2007). *Error handling in Spoken Dialogue Systems*. Doctoral Thesis, Royal Technical University (KTH), Stockholm.

Voice Days (2007). Press release, (in German) e. g.
http://www.computerwoche.de/nachrichten/548530/

---

[7]   Meade. http://en.wikipedia.org/wiki/Time_Is_on_My_Side.

Acknowledgements

The editor would like to thank…

Caroline Clemens M.A., Cener Doğan M.A., Dipl.-Psych. Daniela Hermann, Sharon Lundgren BSc, Stefanie Marek M.A., Margit Straub M.A. (alphabetical order)
   …for being the famous and wonderful "T-PADE" team;

Dr.-Ing. Bernt Andrassy
   …for living best practice sharing in cross departmental work;

Dr. Roman Vilimek
   …for the world's most stimulating scientific discussions – be it voice only or multimodal;

Dave Lewis BA (Hons)
   …for patiently correcting my English, and

Dr. Christoph Baumann
   …for the support at Springer's.

Index